Induced Seismicity Potential in ENERGY TECHNOLOGIES

Committee on Induced Seismicity Potential in Energy Technologies

Committee on Earth Resources

Committee on Geological and Geotechnical Engineering

Committee on Seismology and Geodynamics

Board on Earth Sciences and Resources

Division on Earth and Life Studies

NATIONAL RESEARCH COUNCIL
OF THE NATIONAL ACADEMIES

THE NATIONAL ACADEMIES PRESS
Washington, D.C.
www.nap.edu

THE NATIONAL ACADEMIES PRESS • 500 Fifth Street, NW • Washington, DC 20001

NOTICE: The project that is the subject of this report was approved by the Governing Board of the National Research Council, whose members are drawn from the councils of the National Academy of Sciences, the National Academy of Engineering, and the Institute of Medicine. The members of the committee responsible for the report were chosen for their special competences and with regard for appropriate balance.

This study was supported by DE-PI0000010, TO# 10/DE-DT0001995 between the National Academy of Sciences and the Department of Energy. Any opinions, findings, conclusions, or recommendations expressed in this publication are those of the author(s) and do not necessarily reflect the views of the organizations or agencies that provided support for the project.

International Standard Book Number-13: 978-0-309-25367-3
International Standard Book Number-10: 0-309-25367-5

Additional copies of this report are available for sale from the National Academies Press, 500 Fifth Street, NW, Keck 360, Washington, DC 20001; (800) 624-6242 or (202) 334-3313; http://www.nap.edu/.

Front cover: Photo on right-hand side of photo panel is credited to Julie Shemeta; photo used with permission. Background image is courtesy of the U.S. Geological Survey (http://earthquake.usgs.gov/earthquakes/eqarchives/poster/2011/20110228.php). Cover design by Michael Dudzik.

Printed in the United States of America

COMMITTEE ON INDUCED SEISMICITY POTENTIAL IN ENERGY TECHNOLOGIES

MURRAY W. HITZMAN, *Chair*, Colorado School of Mines, Golden
DONALD D. CLARKE, Geological Consultant, Long Beach, California
EMMANUEL DETOURNAY, University of Minnesota, Minneapolis, and CSIRO (Earth Science and Resource Engineering), Australia
JAMES H. DIETERICH, University of California, Riverside
DAVID K. DILLON, David K. Dillon PE, LLC, Centennial, Colorado
SIDNEY J. GREEN, University of Utah, Salt Lake City
ROBERT M. HABIGER, Spectraseis, Denver, Colorado
ROBIN K. MCGUIRE, Lettis Consultants International, Inc., Boulder, Colorado
JAMES K. MITCHELL, Virginia Polytechnic Institute and State University, Blacksburg
JULIE E. SHEMETA, MEQ Geo, Inc., Highlands Ranch, Colorado
JOHN L. (BILL) SMITH, Geothermal Consultant, Santa Rosa, California

National Research Council Staff

ELIZABETH A. EIDE, Study Director
COURTNEY GIBBS, Program Associate
JASON R. ORTEGO, Research Associate
NICHOLAS D. ROGERS, Financial and Research Associate

Preface

Since the 1920s we have recognized that pumping fluids into or out of the Earth has the potential to cause seismic events that can be felt. Seismic events in Basel, Switzerland, between 2006 and 2008 were felt by local residents and were related to geothermal energy development. Strings of small seismic events in Arkansas, Ohio, Oklahoma, and Texas in the past several years have been related to wastewater disposal associated with oil and gas production. These seismic events have brought the issue of induced (human-caused) seismicity firmly into public view.

Ensuring a reliable twenty-first-century energy supply for the United States presents seminal economic, environmental, and social challenges. A variety of conventional and unconventional energy technologies are being developed to meet these challenges, including new technologies associated with shale gas production and geothermal energy. Energy technologies may also produce wastes. "Wastewater" is often produced during oil and gas drilling and is generally managed either by disposal through pumping the fluids back into the subsurface or by storage, treatment, or reuse. Carbon dioxide may also be generated as a by-product of energy production and may be captured and similarly pumped into the ground for storage.

Anticipating public concern about the potential for induced seismicity related to energy development, Senator Bingaman requested that the Department of Energy conduct a study of this issue through the National Research Council. The study was designed to examine the scale, scope, and consequences of seismicity induced during the injection of fluids related to energy production; to identify gaps in knowledge and research needed to advance the understanding of induced seismicity; to identify gaps in induced seismic hazard assessment methodologies and the research needed to close those gaps; and to assess options for interim steps toward best practices with regard to energy development and induced seismicity potential.

The committee (Appendix A) investigated the history and potential for induced seismicity associated with geothermal energy development; with oil and gas production, including enhanced oil recovery and shale gas; and with carbon capture and storage (CCS). The committee examined peer-reviewed literature, documents produced by federal and state agencies, online databases and resources, and information requested from and submitted by external sources. The committee heard from government and industry representatives; from members of the public familiar with the world's largest geothermal operation at The Geysers, California, at a public meeting in Berkeley, California; and from people familiar with shale gas development, enhanced oil recovery, wastewater disposal, and CCS at meetings in

Dallas, Texas, and Irvine, California (Appendix B). Meetings were also held in Washington, D.C., and Denver, Colorado, to explore induced seismicity in theory and in practice.

During the meeting in Northern California, the committee was able to talk with individuals from Anderson Springs and Cobb, California, who live with induced seismicity continuously generated by geothermal energy production. Understanding their concerns and the history of how they have worked with individuals from both industry and local government, together with technical experts from the federal government, to deal with their very tangible issue of induced seismicity brought immediacy to the committee's deliberations. This knowledge was invaluable as the committee explored the concept of a protocol system for responding to induced seismicity with some of the individuals who helped devise the proposed protocol system for induced seismicity caused by or likely related to enhanced geothermal energy development.

This study took place during a period in which a number of small, felt seismic events occurred that had been caused by or were likely related to fluid injection for energy development. Because of their recent occurrence, peer-reviewed publications about most of these events were generally not available. However, knowing that these events and information about them would be anticipated in this report, the committee attempted to identify and seek information from as many sources as possible to gain a sense of the common factual points involved in each instance, as well as the remaining, unanswered questions about these cases. Through this process, the committee has engaged scientists and engineers from academia, industry, and government because each has credible and viable information to add to better understanding of induced seismicity.

This report describes what we know about the potential for induced seismicity related to energy development. It highlights areas where our knowledge is weak and discusses inherent difficulties in dealing with an issue that does not have a well-defined regulatory "home." The committee hopes this report will inform both the public and the decision-making process with respect to an important issue that will undoubtedly become more widely recognized as additional induced seismic events occur.

As chair, I would like to thank the committee members for their dedication and hard work. The committee commends Dr. Elizabeth Eide, the project study director, for helping to make this an exciting learning experience for us all. The committee also benefited from the dedication and excellence of research associate Jason Ortego and program associate Courtney Gibbs.

Murray W. Hitzman, *Chair*
June 2012

Acknowledgments

In addition to its own expertise, the study committee relied on input from numerous external professionals and members of the public with extensive experience in addressing the range of issues related to induced seismicity. These individuals were very generous in sharing their research knowledge from the laboratory and the field, their direct experiences from industry settings and with energy development in the private sector and in government, and their personal experiences in dealing with induced seismic events. We gratefully acknowledge their contributions to help us with this work. In particular, the committee would like to thank the following people: Scott Ausbrooks, Joe Beall, Lisa Block, Jay Braitsch, Mike Bruno, Linda Christian, David Coleman, Tim Conant, Kevin Cunningham, Mark Dellinger, Philip Dellinger, Nancy Dorsey, Ola Eiken, Leo Eisner, Bill Ellsworth, Cheryl Engels, Rob Finley, Cliff Frohlich, Julio Garcia, Domenico Giardini, Jeffrey Gospe, George Guthrie, Craig Hartline, Werner Heigl, Hamilton Hess, Austin Holland, Steve Horton, Ernst Huenges, John Jeffers, Doug Johnson, Don Juckett, Bill Leith, Ernie Majer, Shawn Maxwell, Steve Melzer, Meriel Medrano, Alexander Nagelhout, Jay Nathwani, David Oppenheimer, Susan Petty, Bruce Presgrave, Philip Ringrose, Jim Rutledge, Jean Savy, Alexander Schriener, Serge Shapiro, Karl Urbank, Mark Walters, Charlene Wardlow, Norm Warpinski, Stefan Wiemer, Colin Williams, Melinda Wright, Bob Young, and Mark Zoback.

The helpful assistance we received with regard to planning and executing the field trip and workshop for the committee's meeting in Northern California was also very important. We recognize the contributions from Calpine, the Northern California Power Agency, the Lawrence Berkeley National Laboratory, and the communities of Anderson Springs and Cobb, California, for their excellent cooperation and efforts to provide us with access to necessary information and localities that greatly informed the committee's work.

The committee gratefully acknowledges the support of three standing committees under the Board on Earth Sciences and Resources for their guidance and oversight during the study process: the Committee on Earth Resources, the Committee on Geological and Geotechnical Engineering, and the Committee on Seismology and Geodynamics (Appendix M). This report has been reviewed in draft form by individuals chosen for their diverse perspectives and technical expertise, in accordance with procedures approved by the National Research Council's (NRC's) Report Review Committee. The purpose of this independent review is to provide candid and critical comments that will assist the institution in making its published report as sound as possible and to ensure that the report meets institutional standards for objectivity, evidence, and responsiveness to the study charge. The review comments and draft manuscript remain confidential to protect the integrity of the deliberative process. We wish to thank the following individuals for their participation in the review of this report:

Jon Ake, Nuclear Regulatory Commission, Rockville, Maryland
Dan Arthur, ALL Consulting, Tulsa, Oklahoma
John Bredehoeft, The Hydrodynamics Group, Sausalito, California
Brian Clark, Schlumberger Companies, Sugar Land, Texas
Peter Malin, University of Auckland, New Zealand
W. Allen Marr, Jr., Geocomp Corporation, Acton, Massachusetts
Shawn Maxwell, Schlumberger Canada, Calgary
J. R. Anthony Pearson, Schlumberger Cambridge Research, United Kingdom
Ed Przybylowicz, Eastman Kodak Company (retired), Webster, New York
Carlos Santamarina, Georgia Institute of Technology, Atlanta, Georgia
Mark Zoback, Stanford University, Stanford, California

Although the reviewers listed above provided many constructive comments and suggestions, they were not asked to endorse the conclusions or recommendations nor did they see the final draft of the report before its release. The review of this report was overseen by William L. Fisher, The University of Texas at Austin, and R. Stephen Berry, the University of Chicago, Illinois. Appointed by the NRC, they were responsible for making certain that an independent examination of this report was carried out in accordance with institutional procedures and that all review comments were carefully considered. Responsibility for the final content of this report rests entirely with the authoring committee and the institution.

Contents

EXECUTIVE SUMMARY 1

SUMMARY 5

1 INDUCED SEISMICITY AND ENERGY TECHNOLOGIES, 23
Introduction to Induced Seismicity and Study Background, 23
Earthquakes and Their Measurement, 27
Energy Technologies and Induced Seismicity, 32
Historical Induced Seismicity Related to Energy Activities, 34
Concluding Remarks, 35
References, 35

2 TYPES AND CAUSES OF INDUCED SEISMICITY 37
Introduction, 37
Factors Affecting Initiation and Magnitude of a Seismic Event, 37
Seismicity Induced by Fluid Injection, 46
Seismicity Induced by Fluid Withdrawal, 51
Summary, 56
References, 57

3 ENERGY TECHNOLOGIES: HOW THEY WORK AND THEIR INDUCED SEISMICITY POTENTIAL 59
Geothermal Energy, 59
Conventional Oil and Gas Production Including Enhanced Oil Recovery, 75
Unconventional Oil and Gas Production Including Shale Reservoirs, 83
Injection Wells Used for the Disposal of Water Associated with Energy Extraction, 88
Carbon Capture and Storage, 94
Discussion, 103
References, 111

4 GOVERNMENTAL ROLES AND RESPONSIBILITIES RELATED TO UNDERGROUND INJECTION AND INDUCED SEISMICITY 117
Federal Authorities, 118
State Efforts, 129
Existing Regulatory Framework for Fluid Withdrawal, 135
Concluding Remarks, 136
References, 136

5 PATHS FORWARD TO UNDERSTANDING AND MANAGING INDUCED SEISMICITY IN ENERGY TECHNOLOGY DEVELOPMENT 139
Hazards and Risks Associated with Induced Seismicity, 139
Quantifying Hazard and Risk, 146
References, 150

6 STEPS TOWARD A "BEST PRACTICES" PROTOCOL 151
The Importance of Considering the Adoption of Best Practices, 151
Existing Induced Seismicity Checklists and Protocols, 152
The Use of a Traffic Light Control System, 157
Mitigating the Effects of Induced Seismicity on Public and Private Facilities, 162
References, 164

7 ADDRESSING INDUCED SEISMICITY: FINDINGS, CONCLUSIONS, RESEARCH, AND PROPOSED ACTIONS 165
Types and Causes of Induced Seismicity, 166
Energy Technologies: How They Work, 168
Oversight, Monitoring, and Coordination of Underground Injection Activities for Mitigating Induced Seismicity, 174
Hazards and Risk Assessment, 175
Best Practices, 176

APPENDIXES

A Committee and Staff Biographies 181
B Meeting Agendas 187
C Observations of Induced Seismicity 195
D Letters between Senator Bingaman and Secretary Chu 207
E Earthquake Size Estimates and Negative Earthquake Magnitudes 211

F The Failure of the Baldwin Hills Reservoir Dam 217
G Seismic Event Due to Fluid Injection or Withdrawal 219
H Pore Pressure Induced by Fluid Injection 225
I Hydraulic Fracture Microseismic Monitoring 229
J Hydraulic Fracturing in Eola Field, Garvin County, Oklahoma, and Potential Link to Induced Seismicity 233
K Paradox Valley Unit Saltwater Injection Project 239
L Estimated Injected Fluid Volumes 243
M Additional Acknowledgments 247

Executive Summary

Earthquakes attributable to human activities are called induced seismic events or induced earthquakes. In the past several years induced seismic events related to energy development projects have drawn heightened public attention. Although only a very small fraction of injection and extraction activities at hundreds of thousands of energy development sites in the United States have induced seismicity at levels that are noticeable to the public, seismic events caused by or likely related to energy development have been measured and felt in Alabama, Arkansas, California, Colorado, Illinois, Louisiana, Mississippi, Nebraska, Nevada, New Mexico, Ohio, Oklahoma, and Texas.

Anticipating public concern about the potential for energy development projects to induce seismicity, the U.S. Congress directed the U.S. Department of Energy to request that the National Research Council examine the scale, scope, and consequences of seismicity induced during fluid injection and withdrawal activities related to geothermal energy development, oil and gas development including shale gas recovery, and carbon capture and storage (CCS). The study was also to identify gaps in knowledge and research needed to advance the understanding of induced seismicity; identify gaps in induced seismic hazard assessment methodologies and the research to close those gaps; and assess options for steps toward best practices with regard to energy development and induced seismicity potential.

Three major findings emerged from the study:

1. The process of hydraulic fracturing a well as presently implemented for shale gas recovery does not pose a high risk for inducing felt seismic events.
2. Injection for disposal of wastewater derived from energy technologies into the subsurface does pose some risk for induced seismicity, but very few events have been documented over the past several decades relative to the large number of disposal wells in operation.
3. CCS, due to the large net volumes of injected fluids, may have potential for inducing larger seismic events.

Induced seismicity associated with fluid injection or withdrawal is caused in most cases by change in pore fluid pressure and/or change in stress in the subsurface in the presence of faults with specific properties and orientations and a critical state of stress in the rocks. The factor that appears to have the most direct consequence in regard to induced seismicity is the net fluid balance (total balance of fluid introduced into or removed from the subsurface), although additional factors may influence the way fluids affect the subsurface. While the general mechanisms that create induced seismic events are well understood, we are currently unable to accurately predict the magnitude or occurrence of such events due to the lack of comprehensive data on complex natural rock systems and the lack of validated predictive models.

Energy technology projects that are designed to maintain a balance between the amount of fluid being injected and withdrawn, such as most oil and gas development projects, appear to produce fewer seismic events than projects that do not maintain fluid balance. Hydraulic fracturing in a well for shale gas development, which involves injection of fluids to fracture the shale and release the gas up the well, has been confirmed as the cause for small felt seismic events at one location in the world.

Wastewater disposal from oil and gas production, including shale gas recovery, typically involves injection at relatively low pressures into large porous aquifers that are specifically targeted to accommodate large volumes of fluid. The majority of wastewater disposal wells do not pose a hazard for induced seismicity, though there have been induced seismic events with a very limited number of wells. The long-term effects of a significant increase in the number of wastewater disposal wells for induced seismicity are unknown.

Projects that inject or extract large net volumes of fluids over long periods of time such as CCS may have potential for larger induced seismic events, though insufficient information exists to understand this potential because no large-scale CCS projects are yet in operation. Continued research is needed on the potential for induced seismicity in large-scale CCS projects.

Induced seismicity in geothermal projects appears to be related to both net fluid balance considerations and temperature changes produced in the subsurface. Different forms of geothermal resource development appear to have differing potential for producing felt seismic events. High-pressure hydraulic fracturing undertaken in some geothermal projects has caused seismic events that are large enough to be felt. Temperature changes associated with geothermal development of hydrothermal resources have also induced felt seismicity.

Governmental response to induced seismic events has been undertaken by a number of federal and state agencies in a variety of ways. However, with the potential for increased numbers of induced seismic events due to expanding energy development, government agencies and research institutions may not have sufficient resources to address unexpected events. Forward-looking interagency cooperation to address potential induced seismicity is warranted.

Methodologies can be developed for quantitative, probabilistic hazard assessments of induced seismicity risk. Such assessments should be undertaken before operations begin in areas with a known history of felt seismicity and updated in response to observed, potentially induced seismicity. Practices that consider induced seismicity both before and during the actual operation of an energy project can be employed in the development of a "best practices" protocol specific to each energy technology and site location.

Although induced seismic events have not resulted in loss of life or major damage in the United States, their effects have been felt locally, and they raise some concern about additional seismic activity and its consequences in areas where energy development is ongoing or planned. Further research is required to better understand and address the potential risks associated with induced seismicity.

Summary

Although the vast majority of earthquakes that occur in the world each year have natural causes, some of these earthquakes and a number of lesser-magnitude seismic events are related to human activities and are called induced seismic events or induced earthquakes. Induced seismic activity has been documented since at least the 1920s and has been attributed to a range of human activities, including the impoundment of large reservoirs behind dams, controlled explosions related to mining or construction, and underground nuclear tests. In addition, energy technologies that involve injection or withdrawal of fluids from the subsurface can also create induced seismic events that can be measured and felt. Historically known induced seismicity has generally been small in both magnitude and intensity of ground shaking.

Recently, several induced seismic events related to energy technology development projects in the United States have drawn heightened public attention. Although none of these events resulted in loss of life or significant structural damage, their effects were felt by local residents, some of whom also experienced minor property damage. Particularly in areas where tectonic (natural) seismic activity is uncommon and energy development is ongoing, these induced seismic events, though small in scale, can be disturbing to the public and raise concern about increased seismic activity and its potential consequences.

This report addresses induced seismicity that may be related to four energy development technologies that involve fluid injection or withdrawal: geothermal energy, conventional oil and gas development including enhanced oil recovery (EOR), shale gas recovery, and carbon capture and storage (CCS). These broad categories of energy technologies, including underground wastewater disposal, are discussed in detail as they relate to induced seismic events. The study arose through a request by Senator Bingaman of New Mexico to Department of Energy (DOE) Secretary Stephen Chu. The DOE was asked to engage the National Research Council to examine the scale, scope, and consequences of seismicity induced during the injection of fluids related to energy production; to identify gaps in knowledge and research needed to advance the understanding of induced seismicity; to identify gaps in induced seismic hazard assessment methodologies and the research needed to close those gaps; and to assess options for interim steps toward best practices with regard to energy development and induced seismicity potential. The report responds to this charge and aims to provide an understanding of the nature and scale of induced seismicity caused by or likely related to energy development and guidance as to how best to proceed with safe development of these technologies while minimizing their potential to induce earthquakes that can be felt by the public.

INDUCED SEISMICITY RELATED TO FLUID INJECTION OR WITHDRAWAL AND CAUSAL MECHANISMS

Seismic events have been measured and felt at a limited number of energy development sites in the United States. Seismic events caused by or likely related to energy development have been documented in Alabama, Arkansas, California, Colorado, Illinois, Louisiana, Mississippi, Nebraska, Nevada, New Mexico, Ohio, Oklahoma, and Texas (Figure S.1). Proving that a particular seismic event was caused by human activity is often difficult because conclusions about the causal relationship rely on local data, prior seismicity, and the preponderance of scientific studies. In this report we give examples of seismic events that

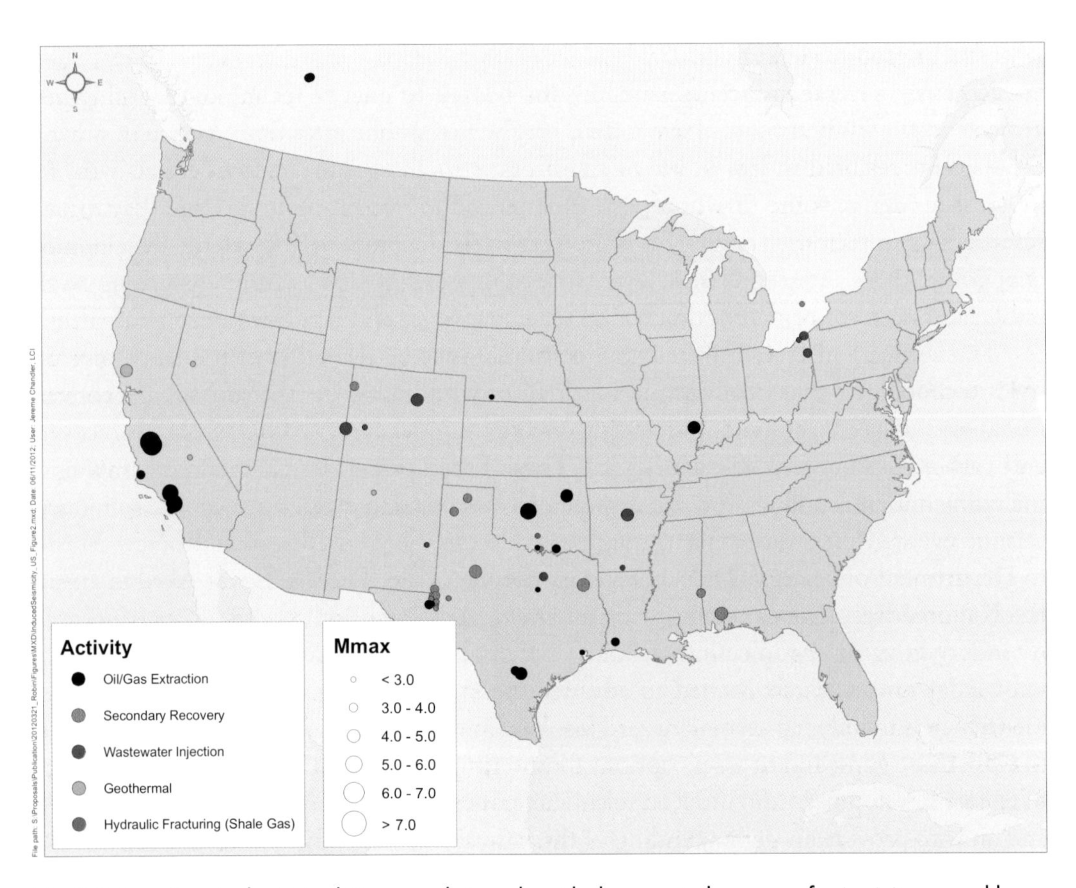

FIGURE S.1 Sites in the United States and Canada with documented reports of seismicity caused by or likely related to energy development from various energy technologies. The reporting of the occurrence of small induced seismic events is limited by the detection and location thresholds of local surface-based seismic monitoring networks.

are universally believed to have been caused by human activities, as well as seismic events for which the evidence for causality is credible but less solid.

Research conducted on some of these incidents has led to better understanding of the probable physical mechanisms of inducing seismic events and allowed for the identification of criteria that could be used to predict whether future induced seismic events might occur. The most important criteria include the amplitude and direction of the state of stress in the Earth's crust in the vicinity of the fluid injection or withdrawal area; the presence, orientation, and physical properties of nearby faults; pore fluid pressure (pressure of fluids in the pores of the rocks at depth, hereafter simply called pore pressure); pore pressure change; the rates and volumes of fluid being injected or withdrawn; and the rock properties in the subsurface.

Seismicity induced by human activity related to energy technologies is caused by change in pore pressure and/or change in stress taking place in the presence of (1) faults with specific properties and orientations and (2) a critical state of stress in the rocks. In general, existing faults and fractures are stable (or are not sliding) under the natural horizontal and vertical stresses acting on subsurface rocks. However, the crustal stress in any given area is perpetually in a state in which any stress change, for example, through a change in subsurface pore pressure due to injecting or extracting fluid from a well, may change the stress acting on a nearby fault. This change in stress may result in slip or movement along that fault, creating a seismic event. Abrupt or nearly instantaneous slip along a fault releases energy in the form of energy waves ("seismic waves") that travel through the Earth and can be recorded and used to infer characteristics of energy release on the fault. Magnitude "**M**" measures the total amount of energy released at the seismic event source, whereas "intensity" of a seismic event is a measure of the level of ground shaking at any location. Both the magnitude and the maximum intensity of a seismic event are directly related to the total area of the fault that undergoes movement: a larger area of slip along the fault results in a larger seismic event.

Although the general mechanisms that create induced seismic events are well understood, current computer modeling techniques cannot fully address the complexities of natural rock systems in large part because the models generally lack information on local crustal stress, rock properties, fault locations and properties, and the shape and size of the reservoir into which fluids are injected or withdrawn. When adequate knowledge of this information is available, the possibility exists to make accurate predictions of earthquake occurrences. Without this detailed information, hazard and risk assessments have to be based on statistical analysis of data from analogous regions. The ability to predict induced seismicity at a particular energy development site will continue to rely on both theoretical modeling and available data including field measurements, and on statistical methods.

ENERGY TECHNOLOGIES AND THEIR INDUCED SEISMICITY POTENTIAL

Geothermal energy, oil and gas production (including hydraulic fracturing for shale gas production), and CCS technologies each involve fluid injection and/or withdrawal. Therefore, each technology has the potential to induce seismic events that can be felt. Seismic events with **M** greater than 2.0 have the possibility of being felt, particularly if they occur at shallow depths, but smaller seismic events (**M** < 2.0) generally are not felt. The injection rate and pressure, fluid volumes, and injection duration vary with the technology as do the potential sizes of the seismic events and the possible risk and hazards of the induced events (Table S.1).

Geothermal Energy

The three different types of geothermal energy resources are (1) "vapor dominated," where primarily steam is contained in the pores or fractures of hot rock; (2) "liquid dominated," where primarily hot water is contained in the rock; and (3) "enhanced geothermal systems" (EGS), where the resource is hot, dry rock that requires engineered stimulation to allow fluid movement for commercial development. Although felt induced seismicity has been documented with all three types of geothermal resources (Table S.1), geothermal development usually attempts to keep a mass balance between fluid volumes produced and fluids replaced by injection to extend the longevity of the energy resource. This fluid balance helps to maintain fairly constant reservoir pressure—close to the initial, preproduction value—and can aid in reducing the potential for induced seismicity. Seismic monitoring at liquid-dominated geothermal fields in the western United States has demonstrated relatively few occurrences of felt induced seismicity. However, in The Geysers geothermal steam field in Northern California, the large temperature difference between the injected fluid and the geothermal reservoir results in significant cooling of the hot subsurface reservoir rocks, causing the rocks to contract, reducing confining pressures, and allowing the release of local stresses that results in a significant amount of observed induced seismicity. EGS technology is in the early stages of development; many countries, including the United States, have pilot projects to test the potential for commercial production. In each case of active EGS development, at least some, generally minor, levels of felt induced seismicity have been recorded.

Conventional and Unconventional Oil and Gas Development, Including EOR and Shale Gas

In a conventional oil or gas reservoir the hydrocarbon fluids and associated aqueous fluids in the pore spaces of the rock are usually under significant natural pressure. Fluids in the oil or gas reservoir flow to the surface when penetrated by a well bore, generally aided by pumping. Oil or gas reservoirs often reach a point when insufficient pressure, even in the presence of pumping, exists to allow sufficient hydrocarbon recovery. Various technologies, including secondary recovery and tertiary recovery (the latter is often referred to as enhanced oil recovery [EOR], which is the term used hereafter), can be used to extract some of the remaining oil and gas. Secondary recovery and EOR technologies both involve injection of fluids into the subsurface to push more of the trapped hydrocarbons out of the pore spaces in the reservoir and to maintain reservoir pore pressure. Secondary recovery often uses water injection or "waterflooding," and EOR technologies often inject carbon dioxide (CO_2). Approximately 151,000 injection wells are currently permitted in the United States for a combination of secondary recovery, EOR, and wastewater disposal, with only very few documented incidents where the injection caused or was likely related to felt seismic events (Table S.1). Secondary recovery—through waterflooding—has been associated with very few felt induced seismic events (Table S.1). Among the tens of thousands of wells used for EOR in the United States, the committee did not find any documentation in the published literature of felt induced seismicity, nor were any instances raised by experts in the field with whom the committee communicated during the study. Oil and gas extraction (fluid withdrawal) from a reservoir may cause induced seismic events. These events are rare relative to the large number of oil and gas fields around the world and appear to be related to a decrease in pore pressure as fluid is withdrawn (Table S.1).

Similar to geothermal systems, conventional oil and gas projects are designed to maintain the pore pressure within a field at its preproduction level by maintaining a balance between fluids being removed from one part of the reservoir and fluids injected in another part of the reservoir. The proportionally very small number of induced seismic events generated by these technologies relative to the large number of wells is in part due to this effort to maintain the original pore pressure of the reservoir.

Shale formations can also contain hydrocarbons—gas and/or oil. The extremely low permeability of these rocks has trapped the hydrocarbons as they developed in the rock and largely prevented them from migrating out of the rock over geologic time. The low permeability also prevents the hydrocarbons from easily flowing into a well bore without production stimulation by the operator. These types of "unconventional" reservoirs are developed by drilling wells horizontally through the reservoir rock and using hydraulic fracturing techniques to create new fractures in the reservoir to allow the hydrocarbons to migrate up the well bore. About 35,000 hydraulically fractured shale gas wells exist in the United States (Table S.1); only one case of felt seismicity (**M** ~ 2.8) in the United States has been described in which hydraulic fracturing for shale gas development is suspected, but not confirmed, as the cause (Table S.1). Globally only one case of felt induced seismicity in England (**M** 2.3) has been confirmed to have been caused by hydraulic fracturing for shale gas development. The very low number of felt events relative to the large number of

TABLE S.1 Summary Information about Historical Felt Seismic Events Caused by or Likely Related to Energy Technology Development in the United States[a]

Energy Technology	Number of Projects	Number of Felt Induced Events	Maximum Magnitude of Felt Events	Number of Events $M \geq 4.0$[b]	Net Reservoir Pressure Change	Mechanism for Induced Seismicity	Location of $M \geq 2.0$ Events
Vapor-dominated geothermal	1	300-400 per year since 2005	4.6	1-3 per year	Attempt to maintain balance	Temperature change between injectate and reservoir	CA (The Geysers)
Liquid-dominated geothermal	23	10-40 per year	4.1[c]	Possibly one	Attempt to maintain balance	Pore pressure increase	CA
Enhanced geothermal systems	~8 pilot projects	2-5 per year[d]	2.6	0	Attempt to maintain balance	Pore pressure increase and cooling	CA, NV
Secondary oil and gas recovery (waterflooding)	~108,000 (wells)	One or more events at 18 sites across the country	4.9	3	Attempt to maintain balance	Pore pressure increase	AL, CA, CO, MS, OK, TX
Tertiary oil and gas recovery (EOR)	~13,000 (wells)	None known	None known	0	Attempt to maintain balance	Pore pressure increase (likely mechanism)	None known
Hydraulic fracturing for shale gas production	35,000 (wells)	1	2.8	0	Initial positive; then withdraw	Pore pressure increase	OK

Hydrocarbon withdrawal	~6,000 fields	20 sites	6.5	5	Withdrawal	Pore pressure decrease	CA, IL, NB, OK, TX
Wastewater disposal wells	~30,000	9	4.8[e]	7	Addition	Pore pressure increase	AR, CO, OH
Carbon capture and storage, small scale	2[f]	None known	None known	0	Addition	Pore pressure increase	IL, MS
Carbon capture and storage, large scale	0	None	None	0	Addition	Pore pressure increase	None yet in operation

[a]Note that in several cases the causal relationship between the technology and the event was suspected but not confirmed. Determining whether a particular earthquake was caused by human activity is often very difficult. The references for the events in this table and the way in which causality may be determined are discussed in the report. **Also important is the fact that the well numbers are those wells in operation today, while the numbers of seismic events that are listed refer to events that have taken place over a total period of decades**.

[b]Although seismic events **M** > 2.0 can be felt by some people in the vicinity of the event, events **M** ≥ 4.0 can be felt by most people and may be accompanied by more significant ground shaking, potentially causing greater public concern.

[c]One event of **M** 4.1 was recorded at Coso, but the committee did not obtain enough information to determine whether or not the event was induced.

[d]Estimate based on the fact that there have been events reported in the mid **M** 2 range at previously active sites and currently active sites but without a large number of total projects (sites) from which to acquire information over time.

[e]**M** 4.8 is a moment magnitude. Earlier studies reported magnitudes up to **M** 5.3 on an unspecified scale; those magnitudes were derived from local instruments.

[f]Noncommercial, pilot projects with active injection supported by the Department of Energy.

hydraulically fractured wells for shale gas is likely due to the short duration of injection of fluids and the limited fluid volumes used in a small spatial area.

Wastewater Disposal Wells Associated with Energy Extraction

In addition to fluid injection directly related to energy development, injection wells drilled to dispose of wastewater generated during oil and gas production are very common in the United States. Tens of thousands of wastewater disposal wells are currently active throughout the country. Although only a few induced seismic events have been linked to these disposal wells (Table S.1), the occurrence of these events has generated considerable public concern. Examination of these cases has suggested causal links between the injection zones and previously unrecognized faults in the subsurface.

In contrast to wells for EOR, which are sited and drilled for precise injection into well-characterized oil and gas reservoirs, injection wells used only for the purpose of wastewater disposal normally do not have a detailed geologic review performed prior to injection, and the data are often not available to make such a detailed review. Thus, the location of possible nearby faults is often not a standard part of siting and drilling these disposal wells. In addition, the presence of a fault does not necessarily imply an increased potential for induced seismicity, creating challenges for the evaluation of potential sites for disposal injection wells that will minimize the possibility for induced seismic activity.

Carbon Capture and Storage

For several years researchers have explored various methods for reducing carbon emissions to the atmosphere, such as by capturing CO_2 and developing means for storing (or sequestering) it permanently underground. If technically successful and economical, CCS could become an important technology for reducing CO_2 emissions to the atmosphere. The risk of induced seismicity from CCS is currently difficult to accurately assess. With only a few small-scale commercial projects overseas and several small-scale demonstration projects under way in the United States, few data are available to evaluate the induced seismicity potential of this technology (Table S.1); these projects so far have involved very small injection volumes. CCS differs from other energy technologies in that it involves continuous CO_2 injection at high rates under pressure for long periods of time, and it is purposely intended for permanent storage (no fluid withdrawal). Given that the potential magnitude of an induced seismic event correlates strongly with the fault rupture area, which in turn relates to the magnitude of pore pressure change and the rock volume in which it exists, large-scale CCS may have the potential for causing significant induced seismicity. CCS projects that do not cause a significant increase in pore pressure above its original value will likely minimize the potential for inducing seismic events.

Energy Technology Summary

The balance of injection and withdrawal of fluids is critical to understanding the potential for induced seismicity with respect to energy technology development projects. The factors important for understanding the potential to generate felt seismic events are complex and interrelated and include the rate of injection or extraction, the volume and temperature of injected or extracted fluids, the pore pressure, the permeability of the relevant geologic layers, faults and fault properties, crustal stress conditions, the distance from the injection point, and the length of time over which injection and/or withdrawal takes place. However, the net fluid balance (total balance of fluid introduced and removed) appears to have the most direct consequence on changing pore pressure in the subsurface over time. Energy technology projects that are designed to maintain a balance between the amount of fluid being injected and the amount of fluid being withdrawn, such as geothermal and most oil and gas development, may produce fewer induced seismic events than technologies that do not maintain fluid balance.

Of the well-documented cases of induced seismicity related to fluid injection, many are associated with operations involving large amounts of fluid injection over significant periods of time. Most wastewater disposal wells typically involve injection at relatively low pressures into large porous aquifers that have high natural permeability and are specifically targeted to accommodate large volumes of fluid. Thus, although a few occurrences of induced seismic activity have been documented, the majority of the hazardous and nonhazardous wastewater disposal wells do not pose a hazard for induced seismicity. However, the long-term effects of any significant increases in the number of wastewater disposal wells on induced seismicity are unknown.

The largest induced seismic events reported in the technical literature are associated with projects that did not balance the large volumes of fluids injected into, or extracted from, the Earth within the reservoir. This is a statistical observation; the net volume of fluid that is injected and/or extracted may serve as a proxy for changes in subsurface stress conditions and pore pressure, injection and extraction rates, and other factors. Coupled thermomechanical and chemomechanical effects may also play a role in changing subsurface stress conditions. Projects with large net volumes of injected or extracted fluids over long periods of time such as long-term wastewater disposal wells and CCS would appear to have a higher potential for larger induced events. The magnitude and intensity of possible induced events would be dependent upon the physical conditions in the subsurface—the state of stress in the rocks, presence of existing faults, fault properties, and pore pressure. The relationship between induced seismicity and projects with large-volume, long-term injection, such as in large-scale CCS projects, is untested because no large-scale projects are yet in existence.

UNDERSTANDING AND MANAGING HAZARDS AND RISKS ASSOCIATED WITH INDUCED SEISMICITY FROM ENERGY DEVELOPMENT

The *hazard of induced seismicity* is the description and possible quantification of what physical effects will be generated by human activities associated with subsurface energy production or CCS. The *risk of induced seismicity* is the description and possible quantification of how induced seismic events might cause losses, including damage to structures, and effects on humans, including injuries and deaths. If seismic events occur in an area with no structures or humans present, there is no risk. The concept of risk can also be extended to include frequent occurrence of ground shaking that is a nuisance to humans.

Several questions can be addressed to understand and possibly quantify the hazard and risk associated with induced seismicity associated with energy technologies. Questions associated with understanding the hazard include whether an energy technology generates apparent seismic events, whether such events are of **M** > 2.0, whether the events generate ground shaking (shallower earthquakes have greater likelihood of causing felt ground shaking than deep earthquakes), and the effects of the shaking. Risk to structures occurs only if the shaking is minor, moderate, or larger; risk to structures does not occur if the shaking is felt by humans but is not strong enough to damage the structures.

The quantification of hazard and risk requires probability assessments, which may be either statistical (based on data) or analytical (based on scientific and engineering models). These assessments can then be used to establish "best practices" or specific protocols for energy project development. A risk analysis of an entire industry project would include the extent of the spatial distribution of the operation and the multiple structures in the area that an induced seismic event might affect. While the risk of minor, moderate, or heavy damage from induced event shaking may be small from an individual well, a large number of spatially distributed wells may lead to a higher probability of such damage; a risk analysis of an industry operation thus includes the entire spatial distribution of the operation and the structures an earthquake might affect.

Although historical data indicate that induced seismic events have not generally been very large nor have they resulted in significant structural damage, induced seismic events are of concern to affected communities. Practices that consider induced seismicity both before and during the actual operation of an energy project can be employed in the development of a "best practices" protocol specific to each energy technology. The aim of such protocols is to diminish the possibility of a felt seismic event occurring and to mitigate the effects of an event if one should occur. A "traffic light" control system within a protocol can be established to respond to an instance of induced seismicity, allowing for low levels of seismicity, but adding monitoring and mitigation requirements, including the requirement to modify or even cease operations if seismic events are of sufficient intensity to result in a significant concern to public health and safety. The ultimate success of such a protocol is fundamentally tied to the strength of the collaborative relationships and dialogue among operators, regulators, the research community, and the public.

GOVERNMENT ROLES AND RESPONSIBILITIES

Four federal agencies—the Environmental Protection Agency (EPA), the Bureau of Land Management, the U.S. Forest Service, and the U.S. Geological Survey (USGS)—and different state agencies have regulatory oversight, research roles, and/or responsibilities related to different aspects of the underground injection activities that are associated with energy technologies. To date, these various agencies have dealt with induced seismic events with different and localized actions. These efforts to respond to potential induced seismic events have been successful but have been ad hoc in nature. Many events that scientists suspect may be induced are not labeled as such, due to lack of confirmation or evidence that those events were in fact induced by human activity. In areas of low historical seismicity, the national seismic network coverage tends to be sparser than that in more seismically active areas, making it difficult to detect small events and to identify their locations accurately.

ADDRESSING INDUCED SEISMICITY

The primary findings, gaps in knowledge or information, proposed actions, and research recommendations to address induced seismicity potential in energy technologies are presented below. Details specific to each energy technology are elaborated in Chapter 7.

Overarching Issues

Findings

1. The basic mechanisms that can induce seismic events related to energy-related injection and extraction activities are not mysterious and are presently well understood.
2. Only a very small fraction of injection and extraction activities among the hundreds of thousands of energy development wells in the United States have induced seismicity at levels that are noticeable to the public.
3. Models to predict the size and location of earthquakes in response to net fluid injection or withdrawal require calibration from field data. The success of these models is compromised in large part due to the lack of basic data on the interactions among rock, faults, and fluid as a complex system; these data are difficult and expensive to obtain.

4. Increase of pore pressure above ambient value due to injection of fluids and decrease in pore pressure below ambient value due to extraction of fluids have the potential to produce seismic events. For such activities to cause these events, a certain combination of conditions has to exist simultaneously:
   a. Significant change in net pore pressure in a reservoir
   b. A preexisting near-critical state of stress along a fracture or fault that is determined by crustal stresses and the fracture or fault orientation
   c. Fault rock properties supportive of a brittle failure
5. Independent capability exists for geomechanical modeling of pore pressure, temperature, and rock stress changes induced by injection and extraction and for modeling of earthquake sequences given knowledge of stress changes, pore pressure changes, and fault characteristics.
6. The range of scales over which significant responses arise in the Earth with respect to induced seismic events is very wide and challenges the ability of models to simulate and eventually predict observations from the field.

BOX S.1
Research Recommendations

Data Collection—Field and Laboratory

1. Collect, categorize, and evaluate data on potential induced seismic events in the field. High-quality seismic data are central to this effort. Research should identify the key types of data to be collected and the data collection protocol.
2. Conduct research to establish the means of making in situ stress measurements nondestructively.
3. Conduct additional field research on microseisms[a] in natural fracture systems including field-scale observations of the very small events and their native fractures.
4. Conduct focused research on the effect of temperature variations on stressed jointed rock systems. Although of immediate relevance to geothermal energy projects, the results would benefit understanding of induced seismicity in other energy technologies.
5. Conduct research that might clarify the in situ links among injection rate, pressure, and event size.

Instrumentation

1. Conduct research to address the gaps in current knowledge and availability of instrumentation: Such research would allow the geothermal industry, for example, to develop this domestic renewable source more effectively for electricity generation.

Hazard and Risk Assessment

1. Direct research to develop steps for hazard and risk assessment for single energy development projects (as described in Chapter 5, Table 5.2).

Gaps

1. The basic data on fault locations and properties, in situ stresses, fluid pressures, and rock properties are insufficient to implement existing models with accuracy on a site-specific basis.
2. Current predictive models cannot properly quantify or estimate the seismic efficiency and mode of failure; geomechanical deformation can be modeled, but a challenge exists to relate this to number and size of seismic events.

Proposed Actions

The actions proposed to advance understanding of the types and causes of induced seismicity involve research recommendations outlined in Box S.1. These recommendations also have relevance for specific energy technologies and address gaps in present understanding of induced seismicity.

Modeling

1. Identify ways simulation models can be scaled appropriately to make the required predictions of the field observations reported.
2. Conduct focused research to advance development of linked geomechanical and earthquake simulation models that could be utilized to better understand potential induced seismicity and relate this to number and size of seismic events.
3. Use currently available and new geomechanical and earthquake simulation models to identify the most critical geological characteristics, fluid injection or withdrawal parameters, and rock and fault properties controlling induced seismicity.
4. Develop simulation capabilities that integrate existing reservoir modeling capabilities with earthquake simulation modeling for hazard and risk assessment. These models can be refined on a probabilistic basis as more data and observations are gathered and analyzed.
5. Continue to develop capabilities with coupled reservoir fluid flow and geomechanical simulation codes to understand the processes underlying the occurrence of seismicity after geothermal wells have been shut in; the results may also contribute to understanding post-shut-in seismicity in relation to other energy technologies.

---

[a]Microseisms designate seismic events that are not generally felt by humans, and in this report are **M** < 2.

Energy Technologies

FINDINGS

1. Injection pressures and net fluid volumes in energy technologies, such as geothermal energy and oil and gas production, are generally controlled to avoid increasing pore pressure in the reservoir above the initial reservoir pore pressure. These technologies thus appear less problematic in terms of inducing felt seismic events than technologies that result in a significant increase or decrease in net fluid volume.
2. The induced seismic responses to injection or extraction differ in cause and magnitude among each of the three different forms of geothermal resources. Decrease of the temperature of the subsurface rocks caused by injection of cold water in a geothermal field has the potential to produce seismic events.
3. The potential for felt induced seismicity due to secondary recovery and EOR is low.
4. The process of hydraulic fracturing a well as presently implemented for shale gas recovery does not pose a high risk for inducing felt seismic events.
5. The United States currently has approximately 30,000 Class II[1] wastewater disposal wells among a total of 151,000 Class II injection wells (which includes injection wells for both secondary recovery and EOR). Very few felt seismic events have been reported as either caused by or temporally associated with wastewater disposal wells; these events have produced felt earthquakes generally less than **M** 4.0. Reducing injection volumes, rates, and pressures has been successful in decreasing rates of seismicity associated with wastewater injection.
6. The proposed injection volumes of liquid CO_2 in large-scale sequestration projects are much larger than those associated with other energy technologies. There is no experience with fluid injection at these large scales and little data on seismicity associated with CO_2 pilot projects. If the reservoirs behave in a similar manner to oil and gas fields, these large net volumes may have the potential to impact the pore pressure over vast areas. Relative to other energy technologies, such large spatial areas may have potential to increase both the number and the magnitude of seismic events.

PROPOSED ACTION

Because of the lack of experience with large-scale fluid injection for CCS, continued research supported by the federal government is needed on the potential for induced seismicity in large-scale CCS projects (see Box S.1). As part of a continued research effort,

[1] Class II wells are specifically those that address injection of brines and other fluids associated with oil and gas production and hydrocarbons for storage.

collaboration between federal agencies and foreign operators of CCS sites is important to understand induced seismic events and their effects on the CCS operations.

Hazards and Risk Assessment

Finding

Risk assessments depend on methods that implement assessments of hazards, but those methods currently do not exist. The types of information and data required to provide a robust hazard assessment would include

- net pore pressures, in situ stresses, and information on faults;
- background seismicity; and
- gross statistics of induced seismicity and fluid injection or extraction.

Proposed Actions

1. A detailed methodology should be developed for quantitative, probabilistic hazard assessments of induced seismicity risk. The goal in developing this methodology would be to

    - make assessments before operations begin in areas with a known history of felt seismicity and
    - update assessments in response to observed induced seismicity.

2. Data related to fluid injection (well location coordinates, injection depths, injection volumes and pressures, time frames) should be collected by state and federal regulatory authorities in a common format and made publicly accessible (through a coordinating body such as the USGS).
3. In areas of high density of structures and population, regulatory agencies should consider requiring that data to facilitate fault identification for hazard and risk analysis be collected and analyzed before energy operations are initiated.

Best Practices

Finding

The DOE Protocol for EGS is a reasonable model for addressing induced seismicity that can serve as a template for protocol development for other energy technologies.

Gap

No best practice protocol for addressing induced seismicity is generally in place for each energy technology. The committee suggests that best practice protocols be adapted and tailored to each technology to allow continued energy technology development.

Proposed Actions

Protocols for best practice should be developed for each of the energy technologies (secondary recovery and EOR for conventional oil and gas production, shale gas production, CCS) by experts in each field, in coordination with permitting agencies, potentially following the model of the DOE EGS protocol. For all the technologies a "traffic light" system should be employed for future operations. The protocols should be applied to

- the permitting of operations where state agencies have identified areas of high potential for induced seismicity or
- an existing operation that is suspected to have caused an induced seismic event of significant concern to public health and safety.

Simultaneous development of public awareness programs by federal or state agencies in cooperation with industry and the research community could aid the public and local officials in understanding and addressing the risks associated with small-magnitude induced seismic events.

Government Roles and Responsibilities

Findings

1. Induced seismicity may be produced by a number of different energy technologies and may involve either injection or extraction of fluid. However, responsibility for oversight of induced seismicity is dispersed among a number of federal and state agencies.
2. Responses to energy development-related seismic events have been addressed in a variety of manners involving local, state, and federal agencies, and research institutions. These agencies and research institutions may not have resources to address unexpected events, and more events could stress this ad hoc system.
3. Currently EPA has primary regulatory responsibility for fluid injection under the Safe Drinking Water Act, which does not explicitly address induced seismicity.

EPA is addressing the issue of induced seismicity through its current study in consultation with other federal and state agencies.

4. The USGS has the capability and expertise to address monitoring and research associated with induced seismic events. However, the scope of its mission within the seismic hazard assessment program is focused on large-impact, natural earthquakes. Significant new resources would be required if the USGS mission were expanded to include comprehensive monitoring and research on induced seismicity.

Gap

No mechanisms are currently in place for efficient coordination of governmental agency response to seismic events that may have been induced.

Proposed Actions

1. Relevant agencies, including EPA, USGS, and land management agencies, and possibly DOE, and state agencies with authority and relevant expertise (e.g., oil and gas commissions, state geological surveys, state environmental agencies) should develop coordination mechanisms to address induced seismic events.
2. Appropriating authorities for agencies with potential responsibility for induced seismicity should consider resource allocations for responding to induced seismic events in the future.

CHAPTER ONE

Induced Seismicity and Energy Technologies

INTRODUCTION TO INDUCED SEISMICITY AND STUDY BACKGROUND

An earthquake is a shaking of the ground caused by a sudden release of energy within the Earth. Most earthquakes occur because of a natural and rapid shift (or slip) of rocks along geologic faults that release energy built up by relatively slow movements of parts of the Earth's crust. The numerous, sometimes large earthquakes felt historically in California and the earthquake that was felt along much of the East Coast in August 2011 are examples of naturally occurring earthquakes related to Earth's movements along regional faults (see also the section Earthquakes and Their Measurement, this chapter). An average of ~14,450 earthquakes with magnitudes above 4.0 (**M** > 4.0)[1] are measured globally every year. This number increases dramatically—to more than 1.4 million earthquakes annually—when small earthquakes (those with greater than **M** 2.0) are included.[2]

Although the vast majority of earthquakes have natural causes, some earthquakes may also be related to human activities and are called induced seismic events.[3] Induced seismic events are usually small in both magnitude and intensity of shaking (see the section on Earthquakes and Their Measurement later in this chapter). For example, underground nuclear tests, controlled explosions in connection with mining or construction, and the impoundment of large reservoirs behind dams can each result in induced seismicity (Box 1.1). Energy technologies that involve injection or withdrawal of fluids from the subsurface also have the potential to induce seismic events that can be measured and felt (see Kerr, 2012).

The earliest and probably most familiar documented example of an induced seismic event related to fluid injection is the activity that occurred in the Denver, Colorado, area in the 1960s in connection with liquid waste disposal at the Rocky Mountain Arsenal. An injection well at the Arsenal pumping into relatively impermeable crystalline basement

[1] **M** represents magnitude on the moment-magnitude scale, which is described in the section Earthquakes and Their Measurement, this chapter.

[2] See earthquake.usgs.gov/learn/faq/?faqID=69.

[3] Some researchers (e.g., McGarr et al., 2002) draw a distinction between "induced" seismicity and "triggered" seismicity. Under this distinction, induced seismicity results from human-caused stress changes in the Earth's crust that are on the same order as the ambient stress on a fault that causes slip. Triggered seismicity results from stress changes that are a small fraction of the ambient stress on a fault that causes slip. Anthropogenic processes cannot "induce" large and potentially damaging earthquakes, but anthropogenic processes could potentially "trigger" such events. In this report we do not distinguish between the two and use the term "induced seismicity" to cover both categories.

BOX 1.1
Observations of Induced Seismicity

Seismicity induced by human activity has been observed and documented since at least the 1920s (Pratt and Johnson, 1926). The number of sites where seismic events of **M** > 0 have occurred that are caused by or likely related to energy development are listed below by technology. (References for these sites with location and magnitude information are in Appendix C; note that in several cases the causal relationship between the technology and the event was suspected but never confirmed.) The numbers of sites globally are listed first in the column; the world map (Figure 1) shows these sites by technology and magnitude. The numbers in parentheses are the numbers of sites, as a subset of the global totals, in which seismic events in the United States have been caused by or likely related to energy development. In addition to energy technologies that are the topic of this report, the list also shows induced seismicity due to surface water reservoirs (dams) and other activities related to mining.[a] Event locations are plotted on global and U.S. maps in Figures 1 and 2.

	Global (United States only)
Wastewater injection	11 (9)
Oil and gas extraction (withdrawal)	38 (20)
Secondary recovery (water flooding)	27 (18)
Geothermal energy	26 (4)
Hydraulic fracturing (shale gas)	2 (1)
Surface water reservoirs	44 (6)
Other (e.g., coal and solution mining)	8 (3)
Total	156

Note that the figures include locations where a spatial association between seismicity and human activity has suggested a causal relationship, but where a causal relationship has not been positively established. Indeed, establishing such a causal relationship often requires a significant amount of scientific effort and fieldwork in the form of temporary seismometer arrays, particularly for the remote locations at which underground activities are conducted.

[a] Mining operations can cause seismic events, in addition to the explosions that are used to fracture rock for excavation. These seismic events may occur at shallow depths as a result of changes in crustal stress, both by removal of mining ore and by redistribution of crustal stress from fracturing sound rock. Such events are not considered further in this report.

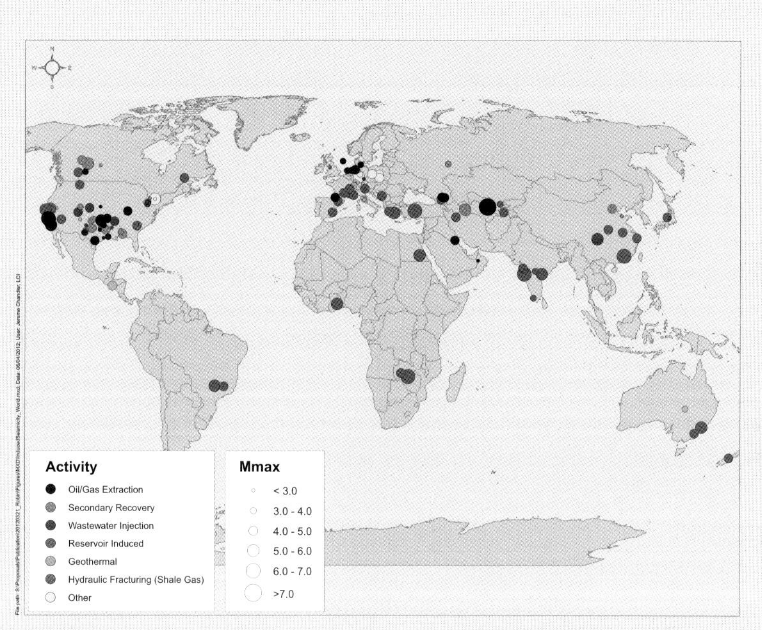

Figure 1 Worldwide locations of seismicity reported in the technical literature caused by or likely related to human activities, with the maximum magnitude reported to be induced at each site.

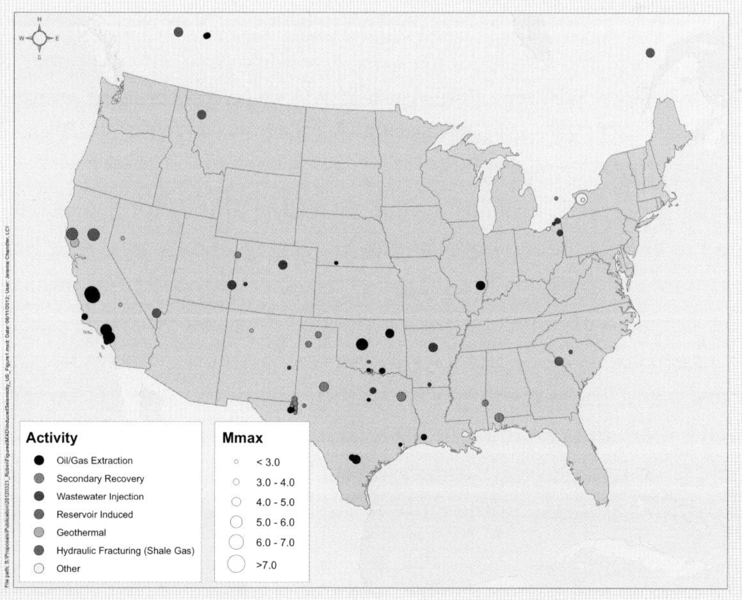

Figure 2 Locations of seismic events caused by or likely related to human activities within the coterminous United States and portions of Canada as documented in the technical literature.

rock caused induced earthquakes (three **M** 5.0 to **M** 5.5 earthquakes[4]), the largest of which caused an estimated $500,000 in damages in 1967 (Nicholson and Wesson, 1990) (Box 1.2).

More recent public attention to the potential correlation between seismic events and energy technology development began with several felt seismic events: in Basel, Switzerland, in 2006; at The Geysers, California, in 2008; and near the Dallas-Fort Worth airport in 2008. During the course of this study, several additional seismic events with potential correlation to energy development have occurred in different parts of the United States and in several other nations (see later in this chapter and in Chapters 2 and 3 for details of some of these events). The potential for induced seismic events has also been highlighted in the context of ongoing public discussion of shale gas development through hydraulic fracturing operations. Although none of these recent events resulted in loss of life or significant structural damage, their effects were felt by local residents, some of whom also experienced minor property damage. Particularly in areas where tectonic (natural) seismic activity is uncommon or historically nonexistent and energy development is ongoing, these seismic events, though small in scale, can be disturbing for the public and can raise concern about further seismic activity and its consequences.

This report addresses induced seismicity that may be related specifically to certain kinds of energy development that involve fluid injection or withdrawal. The study arose through a request made in 2010 by Senator Bingaman of New Mexico, chair of the Senate Energy and Natural Resources Committee, to Department of Energy Secretary Stephen Chu (Appendix D). The senator asked the secretary to engage the National Research Council to examine the scale, scope, and consequences of seismicity induced by energy technologies and specifically associated with four energy technologies: geothermal energy, shale gas,[5] enhanced oil recovery (EOR), and carbon capture and storage (CCS). The study's statement of task is presented in Box 1.3.

The aim of this report is to provide an understanding of the nature and scale of induced seismicity related to energy technologies and to suggest guidance as to how best to proceed with safe development of these technologies in terms of any potential induced seismicity risks. The report begins with an examination of the types and potential causes or mechanisms for induced seismicity (Chapter 2), reviews the four energy technologies that are the subject of the study and the ways they may induce seismic activity (Chapter 3), and discusses government roles and responsibilities related to underground injection and induced seismicity (Chapter 4). Chapter 5 considers the hazard and risk for induced seismicity and identifies some paths for understanding and managing induced seismicity, with steps toward

[4] The initial reports of the magnitudes of the events at the Rocky Mountain Arsenal did not have details about the magnitude scale being used. Subsequent detailed analysis of seismograms (Herrmann et al., 1981) indicated that the magnitudes of the largest earthquakes were actually **M** 4.5 to **M** 4.8, slightly smaller than the initially reported magnitudes. See Box 1.2 for details.

[5] When the committee uses the term "shale gas," it is referring to dry gas, gas, and some liquids.

best practices for mitigating induced seismicity risk in Chapter 6. Chapter 7 contains the report's findings, conclusions, proposed actions, and research recommendations, including identification of information and knowledge gaps and research and monitoring needs. The remainder of this chapter briefly reviews earthquakes and their measurement, introduces the four energy technologies that are the subject of this report, and presents several historical examples of induced seismic activity related to energy development.

The significance of understanding and mitigating the effects of induced seismicity related to energy technologies has been recognized by other groups as well, both internationally and domestically. The International Partnership for Geothermal Technology Working Group on Induced Seismicity[6] under the auspices of the International Energy Agency, for example, has been addressing the issue as it relates specifically to geothermal energy development. International professional societies such as the Society of Petroleum Engineers and the Society of Exploration Geophysicists are coordinating a public technical workshop on the topic.[7] Within the United States, government agencies such as the Department of Energy and U.S. Geological Survey have also been engaged in explicit efforts to understand and address induced seismicity in technology development. The Environmental Protection Agency has been facilitating a National Technical Working Group on Injection Induced Seismicity[8] since mid-2011 and anticipates releasing a report that will contain technical recommendations directed toward minimizing or managing injection-induced seismicity.

EARTHQUAKES AND THEIR MEASUREMENT

The process of earthquake generation is analogous to a rubber band stretched to the breaking point that suddenly snaps and releases the energy stored in the elastic band. Earthquakes result from slip along faults that release tectonic stresses that have grown high enough to exceed a fault's breaking strength. Strain energy is released by the Earth's crust during an earthquake in the form of seismic waves, friction on the causative fault, and, for some earthquakes, crustal elevation changes. Seismic waves can travel great distances; for large earthquakes they can travel around the globe. Ground motions observed at any location are a manifestation of these seismic waves. Seismic waves can be measured in different ways: earthquake **magnitude** is a measure of the size of an earthquake or the amount of energy released at the earthquake source, while earthquake **intensity** is a measure of the level of ground shaking at a specific location. The distinction between earthquake magnitude and intensity is important because intensity of ground shaking determines what

[6] See http://internationalgeothermal.org/; http://www.iea-gia.org/documents/Switzerland_Inducedseismicity_IPGT_IEA_201105031.pdf

[7] See http://www.spe.org/events/12aden/documents/12ADEN_Brochure.pdf

[8] See http://www.gwpc.org/meetings/uic/2012/proceedings/09McKenzie_Susie.pdf; P. Dellinger, presentation to the committee, September 2011.

BOX 1.2
The Rocky Mountain Arsenal Earthquakes

During the spring of 1962 seismological stations in Colorado began recording a number of small earthquakes near Denver. Although Denver had previously been considered to be in an area of low seismicity, between April 1962 and August 1967 over 1,500 earthquakes were recorded at the seismograph station at Bergen Park, Colorado. Some of the earthquakes were noticeable to local residents and exceeded **M** 3 and **M** 4. The earthquakes were eventually attributed to the underground injection of fluid using a deep well drilled on land known as the Rocky Mountain Arsenal approximately 6 miles northeast of downtown Denver.

The Rocky Mountain Arsenal was used by the U.S. Army from 1942 through 1985 for both the manufacture and the disposal of chemical weapons. In 1961 the army drilled a well on the arsenal grounds for the disposal of chemical fluid wastes by underground injection. The well was drilled to a depth of 12,045 feet into Precambrian crystalline rocks (rocks greater than about 700 million years old) beneath the sedimentary rocks of the Denver basin. Fluid injection began in March 1962, and from that time through September 1963, fluid was injected at an average rate of 181,000 gallons per day (gal/day). Injection was stopped in October 1963, but commenced again from August 1964 through April 1965. During this second injection cycle the fluid was not injected under pressure but was fed to the well under gravity flow at a rate of 65,800 gal/day. In April 1965 pressure injection resumed at a rate of 148,000 gal/day. The maximum injection pressure at any time was 72 bars (1,044 pounds per square inch [psi]).[a]

In April and May 1962, two seismological observatories in the Denver area began recording a series of small earthquakes.

> *In June of 1962 several earthquakes occurred which were large enough to be felt by residents and caused considerable concern. By November of 1965 over 700 shocks had been recorded and, although 75 of these had been felt, no damage was reported. . . ."* (McClain, 1970)

Research conducted in the mid-1960s on the deep injection well located on the Arsenal grounds detailed the correlation between the amount of fluid injected into the Arsenal well and the number of Denver earthquakes (Evans, 1966). This research indicated a strong relationship between injection volumes and earthquake frequency (see Figure). More detailed investigation by several local universities and the U.S. Geological Survey (USGS) gave further support to this conclusion. The research showed the majority of the earthquakes had epicenters within 5 miles of the Arsenal's injection well. The depths of the earthquakes varied from 12,140 to 23,000 feet (3,700 to 7,000 meters) below the surface, which is the depth of Precambrian rocks in the area. Research also showed that the epicenters for the earthquakes aligned in a generally northwest-to-southeast direction, similar to the orientation of a system of natural vertical fractures found in the Precambrian rocks in the area.

Although injection into the Arsenal well ceased in February 1966, earthquake activity continued for several more years. The strongest earthquakes actually occurred after injection into the well was discontinued. A detailed analysis of seismograms (Herrmann et al., 1981) indicated seismic moments of the largest earthquakes that can be converted to **M** 4.5 (April 1967), **M** 4.8 (August 1967), and **M** 4.5 (November 1967). These magnitudes are more accurately determined and somewhat smaller than the magnitudes reported in earlier papers on the

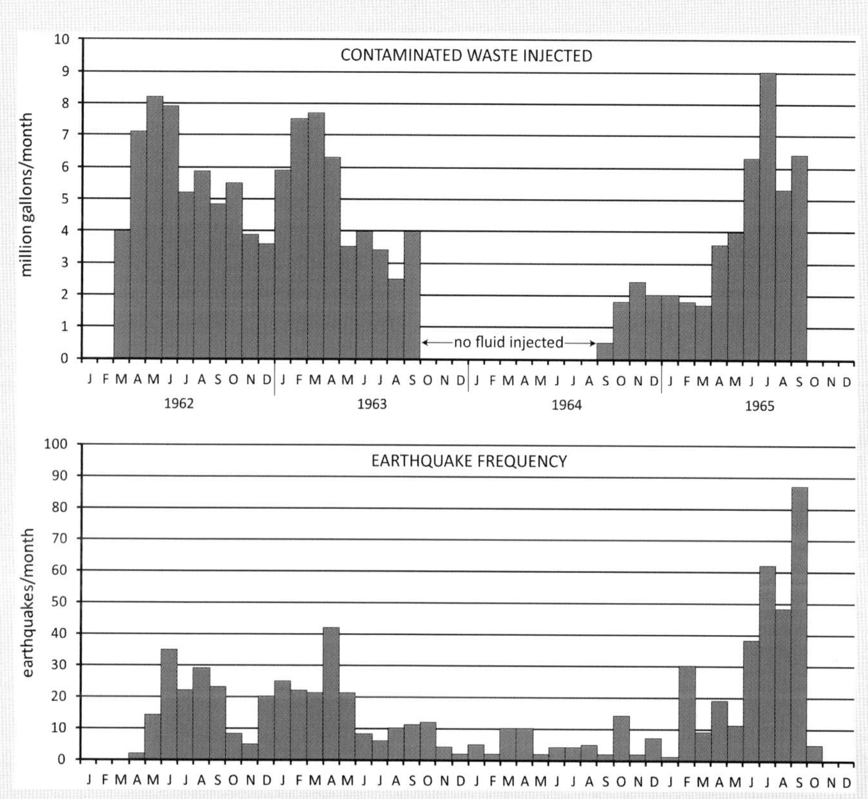

Figure Histograms showing relation between volume of waste injected into the Rocky Mountain Arsenal well and earthquake frequency. SOURCES: Adapted from Evans (1966); Healy et al. (1968); McClain (1970); Hsieh and Bredehoeft (1981).

Rocky Mountain Arsenal earthquakes, which did not have details about the magnitude scale being used. After November 1967 earthquake activity steadily declined and virtually ceased by the late 1980s.

Initial theories postulated that the Denver earthquakes were caused by fluids being pumped into the ground by pressure injection in the disposal well; the fluids were suggested to have acted as a lubricant, allowing large blocks of rock in the subsurface to shift more easily. However, further analysis showed earthquakes triggered by fluid injection are not caused by lubrication of a fracture system but suggested instead that the earthquakes were caused by increasing the pressure of the existing fluid in the formation through high-pressure injection, which lowered the frictional resistance between rocks along an existing fault system; lowering the frictional resistance allowed the rocks to slide relative to each other.

[a] Note: Throughout the report we cite the units presented in the original reference followed by a conversion in parentheses to U.S. measures, metric, or units that might be more familiar to the general reader.

BOX 1.3
Statement of Task

The study will focus on areas of interest related to CCS, enhanced geothermal systems, production from shale gas, and EOR, and will

1. summarize the current state-of-the-art knowledge on the possible scale, scope, and consequences of seismicity induced during the injection of fluids related to energy production, including lessons learned from other causes of induced seismicity;
2. identify gaps in knowledge and the research needed to advance the understanding of induced seismicity, its causes, effects, and associated risks;
3. identify gaps and deficiencies in current hazard assessment methodologies for induced seismicity and research needed to close those gaps; and
4. identify and assess options for interim steps toward best practices, pending resolution of key outstanding research questions.

we, as humans, perceive or feel and the extent of damage to structures and facilities. The intensity of an earthquake depends on factors such as distance from the earthquake source and local geologic conditions, as well as earthquake magnitude. Throughout this work we refer to earthquake magnitudes using the moment-magnitude scale (Hanks and Kanamori, 1979), which is a scale preferred by seismologists because it is theoretically related to the amount of energy released by the Earth's crust. The common symbol used to indicate moment magnitude is **M**.[9]

The earthquake magnitude scale spans a truly immense range of energy releases. For example, an earthquake of **M** 8 does not represent energy release that is four times greater than an earthquake of **M** 2; rather, an **M** 8 releases 792 million times greater energy than an **M** 2. For tectonic ("natural") earthquakes, magnitude is also closely tied to the earthquake rupture area, which is defined as the surface area of the fault affected by sudden slip during an earthquake. A great earthquake of **M** 8 typically has a fault-surface rupture area of 5,000 to 10,000 km^2 (equivalent to ~1,931 to 3,861 square miles or about the size of Delaware,

[9] The moment magnitude scale, designated **M**, is the conventional scale now in use worldwide because it is related to the energy or "work" done by the Earth's crust in creating the earthquake. An earthquake magnitude scale was first published by Richter (1936) and was based on the amplitudes of ground motions recorded on standard seismometers in Southern California. The desire was to assign a numerical magnitude value to earthquakes that was logarithmically proportional to the amount of energy released in the Earth's crust, although it was recognized by Richter that the available data were inadequate for developing a direct correlation with energy. The original scale for Southern California achieved widespread use, was designated "Richter" or local magnitude, and was adapted for other areas with modifications to account for regional differences in earthquake wave attenuation. The moment magnitude has the ability to represent the energy released by very large earthquakes. Moment magnitude, where available, has been used throughout the report.

which is 2,489 square miles). In contrast, **M** 3 earthquakes typically have rupture areas of roughly 0.060 km^2 (about 0.023 square miles or about 15 acres, equivalent to about 15 football fields). "Felt earthquakes" are generally those with **M** between 3 and 5, and "damaging earthquakes" are those with **M** > 5. The maximum velocity of ground shaking is a measure of how damaging the ground motion will be near the fault causing the earthquake. The intensity of shaking at any location is usually expressed using the Modified Mercalli scale and varies from III[10] (felt by few people and would cause hanging objects to sway) for **M** 3, to X (when severe damage would occur). A large earthquake located onshore will generate intensity X near the fault rupture, intensity III at far distances, and all intensities between at intermediate distances.

Most earthquakes, whether natural or induced, that are recorded by seismometers are too small to be noticed by people. These small earthquakes are often referred to as **microearthquakes** or **microseisms**. This report adopts the latter term for all seismic events with magnitude **M** < 2.0. Microseisms as small as **M** –2 (see Appendix E for an explanation of negative magnitudes) are routinely recorded by local **seismometer arrays** during hydraulic fracturing operations used to stimulate oil and gas recovery. At **M** –2 the rupture areas are on the order of 1 m^2 (a little less than 11 square feet).

Most naturally occurring earthquakes occur near the boundaries of the world's tectonic plates where faults are historically active. However, low levels of seismicity also occur *within* the tectonic plates. This fact, together with widespread field measurements of stress and widespread instances of induced seismicity, indicate that the Earth's crust, even in what we may consider geologically or historically stable regions, is commonly stressed near to the critical limit for fault slip (Zoback and Zoback, 1980, 1981, 1989). Because of this natural state of the Earth's crust, *no region can be assumed to be fully immune to the occurrence of earthquakes.*

Induced seismicity may occur whenever conditions in the subsurface are altered in such a way that stresses acting on a preexisting fault reach the breaking point for slip. If stresses in a rock formation are near the critical stress for fault rupture, theory predicts and experience demonstrates that relatively modest changes of pore fluid pressures can induce seismicity. Generally, induced earthquakes are not damaging, but if preexisting stress conditions or the elevated pore fluid pressures are sufficiently high over a large fault area, then earthquakes with enough magnitude or intensity to cause damage can potentially occur.

Identifying whether a particular earthquake or microseism was caused by human activity or occurred naturally is commonly very difficult; often, inferences are made based on spatial and temporal proximity of the earthquake and human activity, on seismic history in the region, and on whether general models of induced seismicity would support a connection. For example, a small amount of fluid injected into the crust at shallow depths (e.g., during

[10] The Mercalli scale uses Roman numerals.

a hydraulic fracturing operation) would not be considered the cause of a **M** 7 earthquake that was initiated at 10 km depth, even if the hydraulic fracturing and earthquake were close in space and time.

The earthquake history of a region also plays a role in inferring whether a particular earthquake was induced. If a certain earthquake appears to be related to human activity, but similar earthquakes have occurred in the past in that region, the connection with human activity is more tenuous than if the correlation between earthquake and human activity occurred in a previously aseismic region. In the latter case, an important indicator might be the *rate* of occurrence of multiple earthquakes, compared to the historical rate (Ellsworth et al., 2012). The important point is that there often is no definitive proof that a particular earthquake was induced; conclusions are usually based on inference.

ENERGY TECHNOLOGIES AND INDUCED SEISMICITY

Geothermal Energy

Geothermal energy production captures the natural heat of the Earth to generate steam that can drive a turbine to produce electricity. Geothermal systems fall into one of three different categories: (1) **vapor-dominated** systems, (2) **liquid-dominated** systems, and (3) **enhanced geothermal systems** (EGS). Vapor-dominated systems are relatively rare. A major example is The Geysers geothermal field in Northern California. Liquid-dominated systems are used for geothermal energy in Alaska, California, Hawaii, Idaho, Nevada, and Utah. In both of these types of hydrothermal resource systems, either steam or hot water is extracted from naturally occurring fractures within the rock in the subsurface and cold fluid is injected into the ground to replenish the fluid supply. EGS are a potentially new source of geothermal power in which the subsurface rocks are naturally hot and fairly impermeable, and contain relatively little fluid. Wells are used to pump cold fluid into the hot rock to gather heat, which is then extracted by pumping the fluid to the surface. In some cases a potential EGS reservoir may lack sufficient connectivity via fractures to allow fluid movement through rock. In this case the reservoir may be fractured using high-pressure fluid injection in order to increase permeability. Permeability is a measure of the ease with which a fluid flows through a rock formation. (See Chapter 2 for detailed discussion of permeability and its relevance to fracture development and fluid flow.) In each of these geothermal systems, the injection or extraction of fluid has the potential to induce seismic activity. Further description of these technologies and examples of induced seismic activity are provided in Chapter 3.

Oil and Gas Production

Oil and gas production involves pumping hydrocarbon liquids (petroleum and natural gas), often together with large amounts of aqueous fluids (groundwater) that commonly contain high amounts of dissolved solids and salts ("brine"), from the subsurface. In the United States, oil and gas operators are required to manage these aqueous fluids through some combination of treatment, storage, disposal, and/or use, subject to government regulations. Commonly, these fluids, if not reused in the extraction process (see also Carbon Capture and Storage, below), are disposed of by injection into the deep subsurface in wells that may be located at some distance from the site of the oil or gas extraction (see also Chapters 3 and 4).

Fluids may also be produced from a well during "flow-back operations" after a well has been hydraulically fractured. Hydraulic fracturing is a method of stimulating an oil- or gas-producing geologic formation by injecting fluid underground to initiate fractures in the rock to aid oil or gas production from the well. A portion of the fluid is later recovered from the well and may be reinjected for additional hydraulic fracture treatments or managed through storage, permanent disposal in an injection well, or treatment for disposal or beneficial use similar to aqueous fluids that are normally produced directly from an oil or gas reservoir. Injection of fluids related to hydraulic fracturing and injection of waste fluids into the subsurface for permanent disposal are two different processes described in detail in Chapter 3.

Oil and gas production (withdrawal) often includes fluid reinjection. The reinjected fluid may be natural gas, aqueous fluids, or carbon dioxide (CO_2) used to help push more oil and gas out from the rocks and to the surface; such reinjection is termed secondary recovery. Enhanced oil recovery, also known as tertiary recovery, uses technologies that also aid in increasing the recovery of hydrocarbons from a reservoir by changing the properties of the oil (primarily aiming to lower the viscosity of the oil so that it flows more easily). The most common EOR techniques involve injecting CO_2 or hydrocarbons, or heating the oil through steam injection or combustion. The injection of fluid to facilitate oil and gas production, similar to fluid injection for geothermal systems, has the potential to generate induced seismic activity. To date, EOR has not been associated with induced seismicity, although felt seismic events have been documented in connection with waterflooding for secondary recovery. The withdrawal of oil and gas has also been associated with induced seismic activity. All of these technologies and examples of induced seismic activity are described further in Chapter 3.

Carbon Capture and Storage

Carbon capture and geologic storage is the separation and capture of CO_2 from emissions of industrial processes, including energy production, and the transport and permanent storage of the CO_2 in deep underground formations. Currently five different types of underground formations are being investigated for permanent CO_2 storage: (1) oil and gas reservoirs, (2) saline formations, (3) unmineable coal seams, (4) organic-rich shales, and (5) basalt formations.[11] Carbon dioxide has been injected into oil and gas reservoirs for several decades to enhance oil recovery. Current large-scale CCS projects in the United States are focused on injection of carbon dioxide into saline brines in regional aquifers. Carbon dioxide must be in the supercritical (liquid) phase to minimize the required underground storage volume; this requires a fluid pressure of greater than 6.9 MPa (about 68 atm[12]) and temperature greater than 31.1°C, which can be achieved at depths greater than about 2,600 feet (~800 meters) (Sminchak et al., 2001). Because no large-scale CCS projects have been completed in the United States, no data or reports on induced seismic activity are available. Chapter 3 reviews in more detail the CCS research and development projects ongoing in the United States, as well as three small, commercial CCS projects overseas.

HISTORICAL INDUCED SEISMICITY RELATED TO ENERGY ACTIVITIES

In the United States, seismicity caused by or likely related to energy development activities involving fluid injection or withdrawal has been documented in Alabama, Arkansas, California, Colorado, Illinois, Louisiana, Mississippi, Nebraska, Nevada, New Mexico, Ohio, Oklahoma, and Texas (see Chapters 2 and 3 for details). Appendix C lists documented and suspected cases globally and in the United States of induced seismicity, including, for example, seismic events caused by waste injection at the Rocky Mountain Arsenal (Healy et al., 1968; Hsieh and Bredehoeft, 1981; Box 1.1) and in the Paradox Basin of western Colorado (see Appendix K); secondary recovery of oil in Colorado (Raleigh et al., 1972), southern Nebraska (Rothe and Lui, 1983), western Texas (Davis, 1985; Davis and Pennington, 1989), and western Alberta (Milne, 1970) and southwestern Ontario, Canada (Mereu et al., 1986); and fluid stimulation to enhance geothermal energy extraction in New Mexico (Pearson, 1981), at The Geysers, California (see Box 3.1), and in Basel, Switzerland (see Box 3.3). Suckale (2010) provides a thorough overview of seismicity induced by hydrocarbon production. Investigations of some of these cases have led to better understanding of the probable physical mechanisms of inducing seismic events and have allowed for the establishment of some of the most important criteria that may induce a felt seismic event, including the state of stress in the Earth's crust in the vicinity of the fluid injection or withdrawal; the presence, orientation, and physical properties of nearby faults; pore fluid pressure (pressure of fluids in the pores of the rocks at depth, hereafter referred to as pore pressure); the volumes, rates, and temperature of fluid being injected or withdrawn; the pressure at which the fluid is being injected; and the length of time over which the fluid is

[11] See, for example, http://www.netl.doe.gov/technologies/carbon_seq/FAQs/carbonstorage2.html.

[12] One unit of atmospheric pressure or 1 atm is equivalent to the pressure exerted by the Earth's atmosphere on a point at sea level.

injected or withdrawn (e.g., Nicholson and Wesson, 1990). Controlled experiments both at Rangely, Colorado (Raleigh et al., 1976; see also Chapter 2), and in Matsushiro, Japan (Ohtake, 1974), were undertaken to directly control the behavior of large numbers of small seismic events by manipulation of fluid injection pressure.

Fluid withdrawal has also been observed to cause seismic events. McGarr (1991) identified three earthquakes in California caused by or likely related to extraction of oil: (1) Coalinga, in May 1983, **M** 6.5; (2) Kettleman North Dome, in August 1985, **M** 6.1; and (3) Whittier Narrows, in October 1987, **M** 5.9. All three events occurred in a crustal anticline close to active oil fields and on or near seismically active faults. Although seismic deformation (uplift) observed during each earthquake has been suggested to have a correlation to removal of hydrocarbon mass (McGarr, 1991), well-documented and ongoing uplift and seismicity over the entire region, related to natural adjustments of the Earth's crust, make it difficult to determine unequivocally if these were induced seismic events. In the mid-1970s and 1980s three large earthquakes (measuring **M** ~ 7) were recorded near the Gazli gas field in Uzbekistan in an area that had largely been aseismic. Although precise locations and magnitudes of the earthquakes were not possible to determine, a potential relation to gas extraction was suggested based on available data and modeling (Adushkin et al., 2000; Grasso, 1992; Simpson and Leith, 1985).

Some surface effects associated with energy technologies may occur (without associated shaking at the surface) that result from surface subsidence or "creep" rather than from slip along a fault. Examples include the Baldwin Hills dam failure in California (Appendix F).

CONCLUDING REMARKS

Human activity, including injection and extraction of fluids from the Earth, can induce seismic events. While the vast majority of these events have intensities below that which can be felt by people living directly at the site of fluid injection or extraction, potential exists to produce significant seismic events that can be felt and cause damage and public concern. Examination of known examples of induced seismicity can aid in determining what the risks are for energy technologies. These examples also provide data on the types of research required to better constrain induced seismicity risks and to develop options for best practices to define and alleviate risks from energy-related induced seismicity. These issues are explored in the remaining chapters of this report.

REFERENCES

Adushkin, V.V., V.N. Rodionov, S.T. Turuntnev, and A.E. Yodin. 2000. Seismicity in the oil field. *Oilfield Review* Summer:2-17.

Davis, S.D. 1985. Investigations of Natural and Induced Seismicity in the Texas Panhandle. M.S. thesis. The University of Texas, Austin. 230 pp.

Davis, S.D., and W.D. Pennington. 1989. Induced seismic deformation in the Cogdell oil field of West Texas. *Bulletin of the Seismological Society of America* 79(5):1477-1494.

Ellsworth, W.L., S.H. Hickman, A.L. LLeons, A. McGarr, A.J. Michael, and J.L. Rubenstein. 2012. Are seismicity rate changes in the Midcontinent natural or manmade? Abstract #12-137 presented at the Seismological Society of America (SSA) 2012 Annual Meeting, San Diego, CA, April 17-19.

Evans, D.M. 1966. The Denver area earthquakes and the Rocky Mountain Arsenal disposal well. *Mountain Geologist* 3(1):23-26.

Grasso, J.-R., 1992. Mechanics of seismic instabilities induced by the recovery of hydrocarbons. *Pure and Applied Geophysics* 139(3-4):507-534.

Hanks, T.C., and H. Kanamori. 1979. A moment magnitude scale. *Journal of Geophysical Research* 84(B5):2348-2350.

Healy, J.H., W.W. Rubey, D.T. Griggs, and C.B. Raleigh. 1968. The Denver earthquakes. *Science* 161:1301-1310.

Herrmann, R. B., S. K. Park, and C. Y. Wang. 1981. The Denver Earthquakes of 1967-1968. *Bulletin of the Seismological Society of America* 71:731-745.

Hsieh, P.A., and J.S. Bredehoeft. 1981. A reservoir analysis of the Denver earthquakes: A case of induced seismicity. *Journal of Geophysical Research* 86(B2):903-920.

Kerr, R. 2012. Learning how to NOT make your own earthquakes. *Science* 335:1436-1437.

McClain, W.C. 1970. On earthquakes induced by underground fluid injection. ORNL-TM-3154. Oak Ridge, TN: Oak Ridge National Laboratory.

McGarr, A. 1991. On a possible connection between three major earthquakes in California and oil production. *Bulletin of the Seismological Society of America* 81(3):948-970.

McGarr, A., D. Simpson, and L. Seeber. 2002. Case histories of induced and triggered seismicity. Pp. 647-661 in *International Handbook of Earthquake and Engineering Seismology, Part A*, edited by W.H.K. Lee et al. New York: Academic Press.

Mereu, R.F., J. Brunei, K. Morrissey, B. Price, and A. Yapp. 1986. A study of the microearthquakes of the Gobies oil field area of southwestern Ontario. *Bulletin of the Seismological Society of America* 76:1215-1223.

Milne, W.G. 1970. The Snipe Lake, Alberta earthquake of March 6, 1970. *Canadian Journal of Earth Science* 7(6):1564-1567.

Nicholson, C., and R.L. Wesson. 1990. Earthquake hazard associated with deep well injection: A report to the U.S. Environmental Protection Agency. U.S. Geological Survey (USGS) Bulletin 1951. Reston, VA: USGS. 74 pp.

Ohtake, M. 1974. Seismic activity induced by water injection at Matsushiro, Japan. *Journal of Physics of the Earth* 22:163-176.

Pearson, C. 1981. The relationship between microseismicity and high pore pressures during hydraulic stimulation experiments in low permeability granitic rocks. *Journal of Geophysical Research* 86(B9):7855-7864.

Pratt, W.E., and D.W. Johnson. 1926. Local subsidence of the Goose Creek oil field (Texas). *Bulletin of the Seismological Society of America* 34(7):577-590.

Raleigh, C.B., J.H. Healy, and J.D. Bredehoeft. 1972. Faulting and crustal stresses at Rangely, Colorado. *American Geophysical Union Geophysics Monograph Series* 16:275-284.

Raleigh, C.B., J.H. Healy, and J.D. Bredehoeft. 1976. An experiment in earthquake control at Rangely, Colorado. *Science* 191(4233):1230-1237.

Richter, C.F. 1936. An instrumental earthquake magnitude scale. *Bulletin of the Seismological Society of America* 25:1-32.

Rothe, G.H., and C.-Y. Lui. 1983. Possibility of induced seismicity in the vicinity of the Sleepy Hollow oil field, southwestern Nebraska. *Bulletin of the Seismological Society of America* 73(5):1357-1367.

Simpson, D.W., and W. Leith. 1985. The 1976 and 1984 Gazli, USSR, Earthquakes—Were they induced? *Bulletin of the Seismological Society of America* 75(5):1465-1468.

Sminchak, J., N. Gupta, C. Byrer, and P. Bergman. 2001. Issues related to seismic activity induced by the injection of CO_2 in deep saline aquifers. Presented at the First National Conference on Carbon Sequestration, Washington, DC, May 15-17.

Suckale, J. 2010. Moderate-to-large seismicity induced by hydrocarbon production. *The Leading Edge* 29(3):310-319.

Zoback, M.D., and M.L. Zoback. 1981. State of stress and intraplate earthquakes in the United States. *Science* 213(4503):96-104.

Zoback, M.L., and M.D. Zoback. 1980. State of stress in the coterminous United States. *Journal of Geophysical Research* 85(B11):6113-6156.

Zoback, M.L., and M.D. Zoback. 1989. Tectonic stress field of the continental United States. *Geophysical Framework of the Continental United States*, edited by L. Pakiser and W. Mooney. *Geological Society of America Memoir* 172:523-539.

CHAPTER TWO

Types and Causes of Induced Seismicity

INTRODUCTION

Energy technology activities known to have produced induced seismicity, whether significant enough to be felt by humans or so small as to be detected only with sensitive monitoring equipment, are fluid injection and withdrawal as well as purposeful fracturing of rocks. For each of these activities the critical components required to produce induced seismicity are the presence and orientation of existing faults, the state of stress of the Earth's crust, the rates and volumes of fluid injection or withdrawal, and time. Understanding these components gives some confidence in being able to draw conclusions about what seismicity might be induced in the future, and under what conditions. The physical mechanisms[1] responsible for inducing seismic events are discussed here with reference to specific energy technologies; detailed explanations of these technologies and their relationship to induced seismic events are presented in Chapter 3.

FACTORS AFFECTING INITIATION AND MAGNITUDE OF A SEISMIC EVENT

Shallow earthquakes result from slip (movement) along a preexisting fault. Two critical questions concerning such earthquakes are (1) which factors are responsible for the initiation of a seismic event and (2) which factors control the magnitude of the event.

Initiation of a Seismic Event

The Earth's crust is crossed by a network of preexisting fractures and faults of various sizes. Any of these faults could, in principle, be activated if the shear stress (τ) acting on the fault overcomes its resistance to slip or movement of the adjacent rock blocks (called "shear resistance"). In most cases, the shear resistance (or shear strength) is due to friction. In other words, the shear strength is proportional to the difference between the normal stress (σ) acting on the fault and the pressure (ρ) of the fluid permeating the fault and the surrounding rock. The fault remains stable (does not slip) as long as the magnitude of

[1] Although hydromechanical coupling is the dominant mechanism responsible for inducing seismic events, other coupling mechanisms (e.g., thermomechanical and chemomechanical) could also play a role.

the shear stress (τ) is smaller than the frictional strength, which can be represented by this expression: $\mu(\sigma - \rho)$. The term $(\sigma - \rho)$ is called the effective stress. The symbol μ represents the friction coefficient, a parameter that varies only in a narrow range, typically between 0.6 and 0.8 for most rock types. This condition for triggering slip, known as the Coulomb criterion, is discussed in more detail in Box 2.1 and Appendix G (see also Jaeger et al., 2007; Scholz, 2002).

The key parameters controlling the initiation of slip are therefore the normal and shear stresses acting on the fault as well as the pore fluid pressure (hereafter simply referred to as "pore pressure"). The normal and shear stresses on the fault depend on the orientation of the fault and on the state of stress in the rock. Due to the weight of the overlying rock and other processes in the Earth's crust, rocks are usually under compression. The compressive normal stress acting on a rock at depth varies with direction; this variation of the normal stress with direction is linked to the shear stresses that are responsible for slip along a fault if the frictional resistance of the fault is overcome. In contrast, for a fluid at rest, the state of stress is hydrostatic: the normal stress is the same in all directions, and it cannot transmit any shear stresses.

The state of effective stress at a point in the Earth involves both the stress tensor and the pore pressure. The stress tensor is described by the vertical stress (σ_v) and the minimum and maximum horizontal stresses (σ_h and σ_H) that act in two orthogonal directions. The direction of σ_H, as well as the relative values of σ_v, σ_h, and σ_H, control the orientation of the fault most likely to slip; three different fault regimes are defined depending on the relative magnitude of σ_v, σ_h, and σ_H (Box 2.2). Once the most critical fault orientation has been identified, the normal and shear stresses acting on the fault can in principle be computed from the state of effective stress.

Determination of the in situ state of stresses in the subsurface is complex and often expensive. Consequently, the information on the in situ stress in the Earth is usually too fragmentary to allow confident estimates of the actual stresses acting on a fault. In most cases the only reliable information available is the magnitude of the vertical stress, as it can simply be estimated from the average density of the overlying rock and the depth. Estimating the general fault types and configurations as well as the broad orientation of the maximum and minimum horizontal stresses at a scale of tens or hundreds of kilometers is also sometimes possible, based on a variety of stress indicators (see also Figure 4 in Box 2.2).

In contrast to the difficulty of determining the maximum and minimum horizontal stresses and their orientations, the undisturbed initial pressure of the fluid permeating the rock and the fractures or faults can usually be reliably estimated from the depth of the rocks, under normally pressurized conditions. Techniques also exist for direct measurement of the pore pressure from a well.

Although the conditions for initiating slip on a preexisting fault are well understood, the difficulty remains to make reliable estimates of the various quantities in the Coulomb criterion. Lacking these estimates, predicting how close or how far the fault system is from instability remains difficult, even if the orientation of the fault is known. This implies that the magnitude of the increase in pore pressure that will cause a known fault to slip cannot generally be calculated. Nonetheless, understanding how different factors contribute to slip initiation is valuable because it provides insight about whether fluid injection or withdrawal may be a stabilizing or a destabilizing factor for a fault (in other words, whether fluid injection or withdrawal causes the difference between the driving shear stress and the shear strength to increase or decrease). Any perturbation in the stress or pore pressure that is associated with an increase of the shear stress magnitude and/or a decrease of the normal stress and/or an increase of the pore pressure could be destabilizing; such a perturbation brings the system closer to critical conditions for failure. A large body of evidence suggests that the state of stress and pore pressure are often not far from the critical conditions where a small destabilizing perturbation of the stress and/or of the pore pressure could cause a critically oriented fault to slip (Zoback and Zoback, 1980, 1989).

Magnitude of a Seismic Event

The moment magnitude scale, designated **M**, is directly related to the amount of crustal energy released during a seismic event (Hanks and Kanamori, 1979). This energy can be thought of as the total force released during the earthquake times the average fault displacement over the fault rupture area (see also the section Earthquakes and Their Measurement in Chapter 1).

Earthquake magnitude is correlated to the area of the rupture surface. Earthquakes with large magnitudes always involve large parts of the Earth's crust, because the large energies being released can only be stored in large volumes of rock, and large rupture areas are necessary to produce large fault displacements. Correlations between **M** and rupture area from observations of historical earthquakes indicate that an increase of 1 magnitude unit implies, on average, an increase by a factor of about 8 in fault rupture area, and a concurrent increase by a factor of about 4½ in rupture displacement (Wells and Coppersmith, 1994). The following examples are typical fault rupture areas and rupture displacements associated with earthquakes of **M** 4 and **M** 5:

	M 4	**M 5**
Fault rupture area:	1.4 km^2 (~0.5 mi^2)	11 km^2 (~4.2 mi^2)
Fault displacement:	1 cm (~0.4 in)	4.5 cm (~1.8 in)

A larger-magnitude earthquake implies both a larger area over which crustal stress is released and a larger displacement on the fault. From the definition of **M**, we can expect that a 1-unit increase in magnitude will be associated with a factor of about 32 larger release

BOX 2.1
Conditions Leading to Seismic Slip on a Fault

Shallow earthquakes result from slip along a preexisting fault. The slip is triggered when the stress acting along the fault exceeds the frictional resistance to sliding. The critical conditions are quantified by the Coulomb criterion, which embodies two fundamental concepts, friction and effective stress. These two concepts can be illustrated by considering the shearing of a split block (Figure 1). The block is subjected to a normal force F_n and a shear force F_s, which can be translated into a normal stress $\sigma = F_n/A$ and the shear stress $\tau = F_s/A$ acting across the joint, with A designating the interface area of the joint. The joint (and possibly also the block if it is porous) is infiltrated by fluid at pressure ρ.

According to the Coulomb criterion, there is no relative movement across the joint, as long as the shear stress (τ) is less than the frictional strength $\mu(\sigma - \rho)$, where μ is the coefficient of friction. The conditions for slip are thus met when $\tau = \mu(\sigma - \rho)$. The term $(\sigma - \rho)$ is called the effective stress; the presence of effective stress in the Coulomb criterion shows that the fluid pressure (ρ) counterbalances the stabilizing effect of the normal stress (σ). The Coulomb criterion indicates that slip can be triggered by a decrease of the normal stress, an increase of the pore pressure, and/or an increase of the shear stress (Figure 1b).

Note that the common concept that "injected fluids cause earthquakes by lubricating underground faults" is not accurate because fluids do not decrease the coefficient of friction, μ. Rather, injected fluids (or extracted fluids) cause earthquakes by changing the stress conditions around faults, bringing these stresses into a condition where driving stresses equal or exceed resistive stresses, thereby promoting slip on the fault.

Within the context of slip on a fault, the normal and shear stresses acting across the fault, σ and τ, can be directly expressed in terms of the vertical stress (σ_v), the horizontal stress (σ_h), and the fault inclination (β) (Figure 2). Prior to injection or extraction of fluid, the initial state is stable because the shear stress (τ_o) is less than the frictional strength $\mu(\sigma_o - \rho_o)$, although the condition could be close to critical. Injection or extraction of fluid could cause changes in the stress and pore pressure such that the critical condition expressed as $\tau = \mu(\sigma - \rho)$ is met (Figure 2b is a graphical representation).

This box describes the simple case of a frictional fault. The more general case of a fault with cohesive-frictional strength is treated in Appendix H.

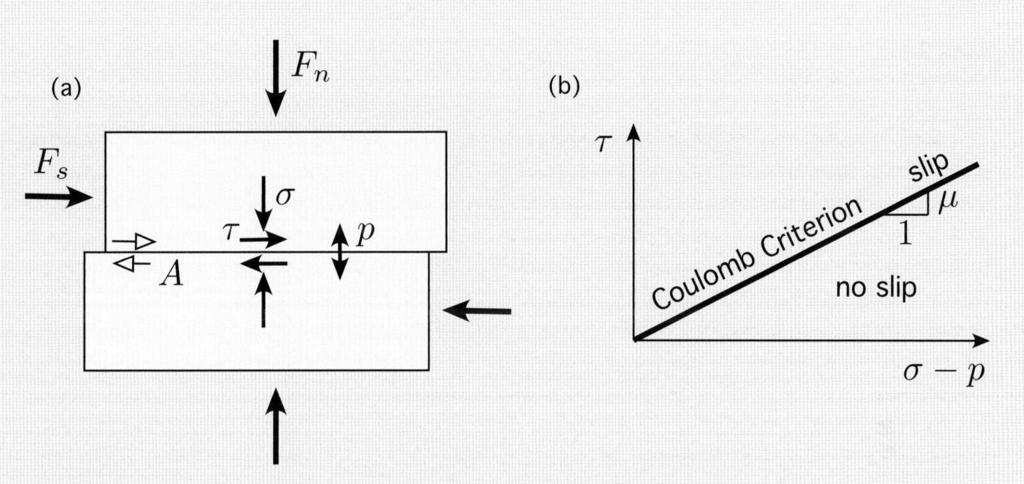

Figure 1 (a) Shearing of a jointed block subjected to normal force F_n and shear force F_s, with fluid inside the joint at pressure ρ. Slip along the joint is triggered when the shear stress τ is equal to the frictional strength μ(σ – ρ), where (σ – ρ) is the effective stress and μ is the coefficient of friction. (b) Graphical representation of the Coulomb criterion: there is no slip if the "point" (σ – ρ, τ) is below the critical line defined by slope μ.

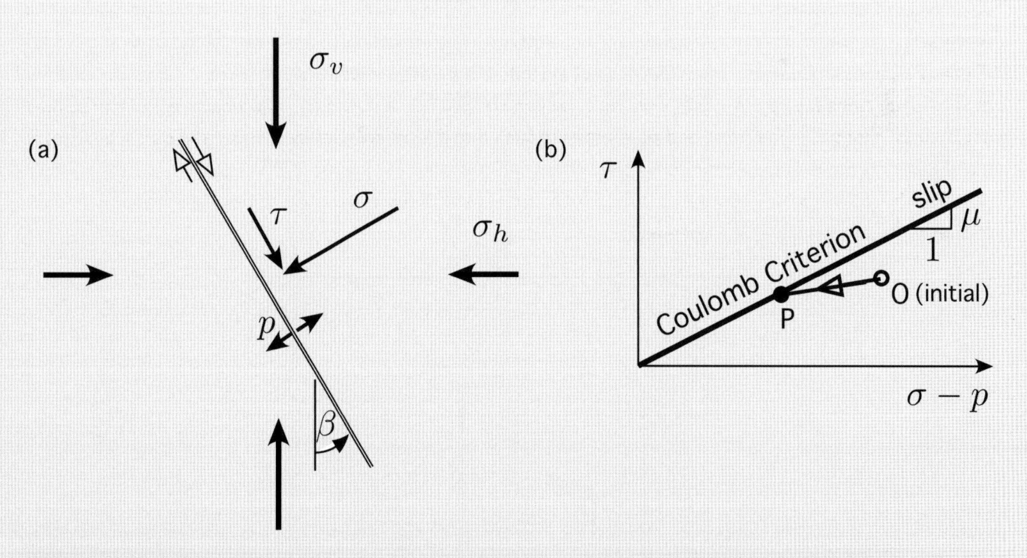

Figure 2 (a) The normal and shear stresses (σ and τ) acting across the fault depend on the vertical and horizontal stress (σ_v and σ_h) and the fault inclination (β). (b) Fluid injection or extraction could induce changes in the stress and the pore pressure; for example, fluid injection could move the initially stable "point" "O" in Figure 2b to a new position "P" that is on the critical Coulomb line, thus triggering slip on the fault. The inclination of the segment OP is a function of the poroelastic coupling described in Box 2.3.

BOX 2.2
In Situ Stress State

The full characterization of the state of stress at a point of the subsurface requires in principle six independent quantities, illustrated through the following example.

Imagine that a small cube of rock centered on the point of interest is cut from its surrounding. To leave the material inside the cube undisturbed by the cutting, forces have to be applied on each face of the cube to mimic the action of the surrounding medium onto the cut material, noting also that forces acting on opposed faces are equal and opposite in direction. However, in considering in situ stress state, using the term "stress," which is equivalent to the force exerted over a defined area, is more appropriate than discussing "force" alone; in this way, stress is not dependent on the size of the cube.

If the cube is rotated in space, the stresses acting on its faces change in magnitude and direction. However, a certain orientation of the cube exists for which each face is only loaded by a stress normal to the face (Figure 1). The three independent normal stresses are referred to as principal stresses, and their corresponding orientations in space as principal directions. On two faces of the cube oriented according to the principal directions, the normal stress is maximum and minimum and for any other orientation of the cube, the normal stress on any face is in between these two limiting values. The principal stress acting on the face parallel to the minimum and maximum principal stresses is called intermediate.

A set of six quantities, the three principal stresses and their directions, thus represents the state of stress. Fortunately, vertical can often be considered as one of the principal directions, with the consequence that the vertical stress σ_v at depth h is then simply given by the weight of the overlying rock (i.e., $\sigma_v = pgh$, where p is the average density of the overlying rock and g is gravity). Determination of the state of stress is then reduced to identifying three quantities, the minimum and maximum horizontal stresses, respectively σ_h and σ_H, and the azimuth of σ_H (or equivalently of σ_h).

Stress data compiled by Brown and Hoek (1978) confirm that, despite some scattering, the vertical stress is proportional to depth in a manner consistent with an average rock density p = 2,700 kg/m^3 (~170 lb/ft^3) (Fig-

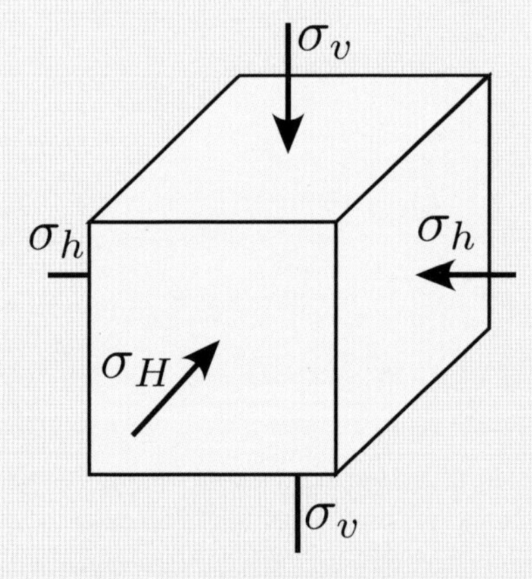

Figure 1 State of stress in the subsurface, with one of the principal stress directions being vertical. By convention, $\sigma_H > \sigma_h$.

ure 2a). The ratio of the mean horizontal stress to the vertical stress (Figure 2b) appears to vary over a narrower range with increasing depth, the ratio being generally less than 1 at depths larger than 2 km (~1.2 miles).

The relative magnitude of the three principal stresses, σ_h , σ_H , and σ_v, establishes the conditions for the orientation of the faults. Three regimes of stress, each associated with different fault orientations, are commonly defined (Figure 3): (a) thrust fault regime with σ_v equal to the minimum principal stress, (b) normal fault regime

continued

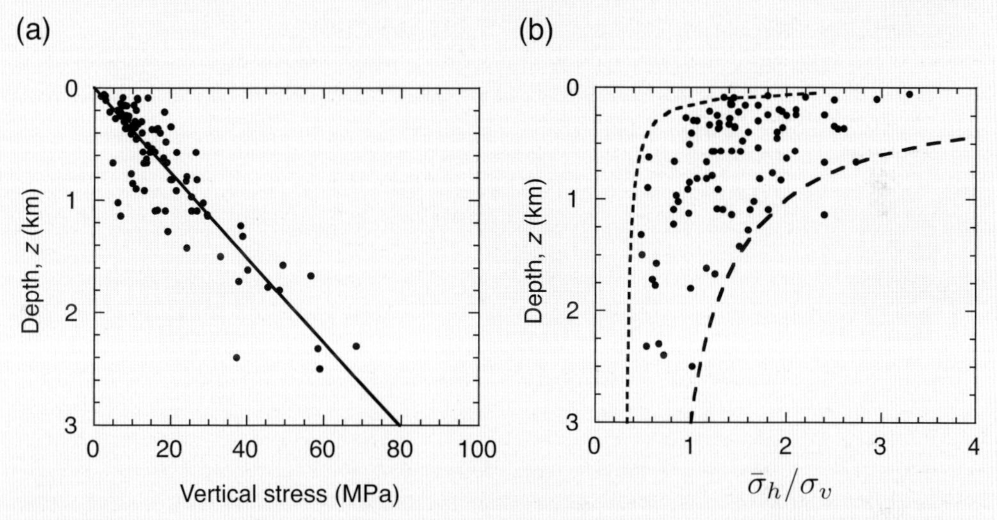

Figure 2 (a) Vertical stress variation with depth; the linear trend corresponds to a mean density of 2,700 kg/m³. (b) Variation of the ratio of the mean horizontal stress $(\sigma_H + \sigma_h)/2$ over the vertical stress σ_v with depth. SOURCES: Figure modified from Jaeger et al. (2007), which was itself redrawn from the original figure of Brown and Hoek (1978).

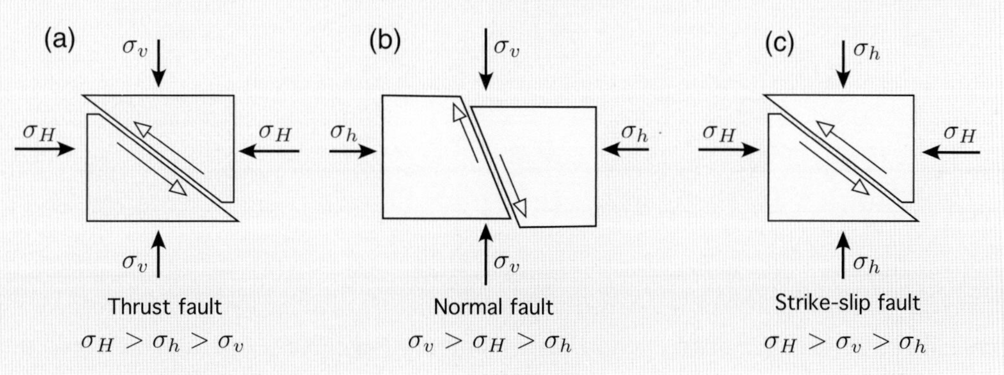

Figure 3 (a) Thrust fault, (b) normal fault, and (c) strike-slip fault. (Cross sections shown are in vertical plane for (a) and (b) and horizontal plane for (c).)

BOX 2.2 Continued

with σ_v equal to the maximum stress, and (c) strike-slip fault regime corresponding to the vertical stress being equal to the intermediate principal stress.

Determination of three unknown quantities (σ_h , σ_H , and their orientation) remains a formidable problem. Most of the time, the only information available is the stress regime and the broad orientation of σ_H , which can be inferred using a variety of stress indicators such as earthquake focal mechanisms, wellbore breakouts, drilling-induced fractures, and other data (Zoback and Zoback, 1980, 1989).

Furthermore, the stress varies from point to point within the Earth, subject to the constraint of having to satisfy the equilibrium equations, a consequence of Newton's second law. Spatial variation of the state of stress exists at various scales, as the stress is affected by the structure of the subsurface, the geometry and mechanical properties of different lithologies, preexisting faults and other discontinuities in the crust, and other characteristics. Yet, when viewed at the scale of hundreds of kilometers, patterns emerge that can be seen on the stress map for North America (Figure 4). This stress map, a compilation of all available stress information, shows the orientation of σ_H and the stress regime superimposed on a topographical map of North America (Heidbach et al., 2008).

The example above refers to the initial state of stress (i.e., to the stress prior to injection or extraction of fluid). Large variation of the pore pressure and/or temperature could also induce significant stress changes that have to be accounted for when assessing the potential for induced seismicity.

in crustal energy (a factor often cited in news reports following large earthquakes), and the estimates cited in the examples from empirical observations are in general agreement with that definition.

Most existing fractures in the Earth's crust are small and capable of generating only

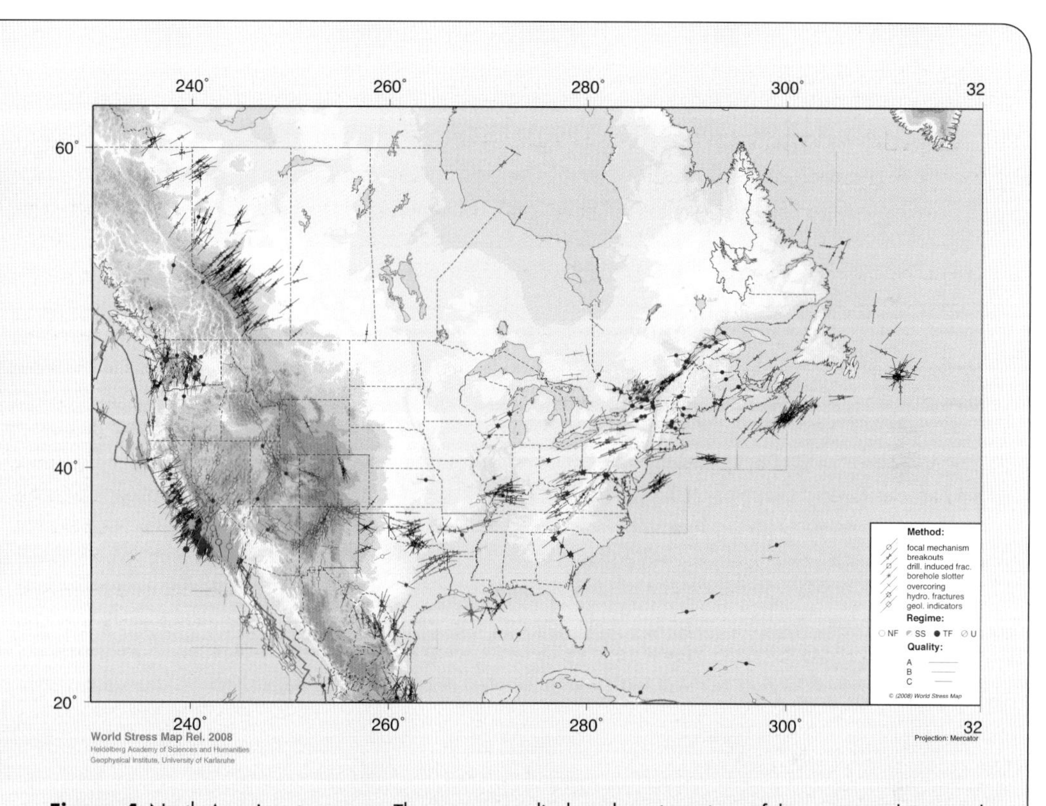

Figure 4 North America stress map. The stress map displays the orientations of the maximum horizontal compressive stress (σ_H). The length of the stress symbols represents the data quality, with A being the best quality. Quality A data are assumed to record the orientation of σ_H to within 10°-15°, quality B data to within 15°-20°, and quality C data to within 25°. As can be seen from this global dataset, stress measurements are absent in many parts of North America and the offshore regions. Because stress measurements are important in the consideration of induced seismicity, their measurement, particularly in areas where data are sparse, could usefully contribute to understanding the potential for induced seismicity related to energy development. The tectonic regimes are NF for normal faulting, SS for strike-slip faulting, TF for thrust faulting, and U for an unknown regime. Topographic relief is indicated by green (lower elevations) to brown (higher elevations) shading. SOURCE: Data used to plot this map were accessed from www.world-stress-map.org/ (see Heidbach et al., 2008).

small earthquakes. Thus, for fluid injection to trigger a significant earthquake, a fault or faults of substantial size must be present that are properly oriented relative to the existing state of crustal stress, and these faults must be sufficiently close to points of fluid injection to have the rocks surrounding them experience a net pore pressure increase.

SEISMICITY INDUCED BY FLUID INJECTION

Injection of fluid in rocks causes an increase of the pore pressure and also modifies the state of the stress (Hsieh, 1996; NRC, 1990). The stress change is associated with a volume expansion of the rock due to the increase of the pore pressure, similar to the familiar thermal expansion experienced by materials (Box 2.3). However, the pore pressure perturbation

BOX 2.3
Stress Induced by Fluid Injection or Withdrawal

Injection or extraction of fluid into or from a permeable rock induces not only a pore pressure change in the reservoir but also a perturbation in the stress field in the reservoir and in the surrounding rock. The physical mechanism responsible for this hydraulically induced stress perturbation can be illustrated by considering the injection of a finite volume of fluid inside a porous elastic sphere surrounded by a large impermeable elastic body (see Figure). The magnitude of the induced pore pressure (Δp), once equilibrated, is proportional to the volume of fluid injected.

Assuming that the sphere is removed from the surrounding body, the pore pressure increase (Δp) induces a free expansion of the sphere (ΔV_*), similar in principle to the familiar thermal expansion experienced by a solid subject in response to a temperature increase. To force the expanded sphere back to its earlier size requires the application of an external confining stress ($\Delta\sigma_*$), which is then relaxed. The final state corresponds to a constrained expansion of the sphere (ΔV), which is less than the free expansion; this state can be associated with a stress perturbation ($\Delta\sigma$) that is isotropic and uniform inside the sphere, but nonisotropic and nonuniform outside the sphere. The magnitude of the stress perturbation decays away from the sphere, becoming negligible at a distance about twice the sphere radius. The stress induced inside the sphere is compressive when the pore pressure increases (fluid injection) but tensile if the pore pressure decreases from its ambient value (fluid withdrawal).

This example illustrates the fundamental mechanism by which the stress field in the rock is modified by injection or withdrawal of fluid. The complexities associated with geological settings—in particular, the actual shape of the reservoir, its size, as well as the nonuniformity of the pore pressure field—affect the nature of the stress perturbation. The horizontal and vertical stress variations within most geological reservoirs are rarely identical; inside a tabular reservoir of large lateral extent compared to its thickness, only the horizontal stress is affected by the pore pressure change.

In the case of fluid injection in a fractured impermeable basement rock, such as that which may be a target for development of enhanced geothermal systems (EGS; see also Chapter 3), the perturbation is only of a hydraulic nature and the stress change can generally be ignored.

An analysis of the pore pressure and stress perturbation indicates that, in general, fluid injection increases the risk of slip along a fault located in the region where the pore pressure has increased. In the case of fluid withdrawal, the region at risk is generally outside the reservoir (see also Nicholson and Wesson, 1990).

dominates over the stress variation and, when the consequence of fluid injection with regard to the induced seismicity is considered, the stress perturbations can often be ignored. Disregarding the stress change in the rock caused by injection is a conservative approach because these kinds of perturbations are usually of a stabilizing nature (see Appendix G for a detailed explanation).

Pore pressure increases in the joints and faults are potentially destabilizing, since they

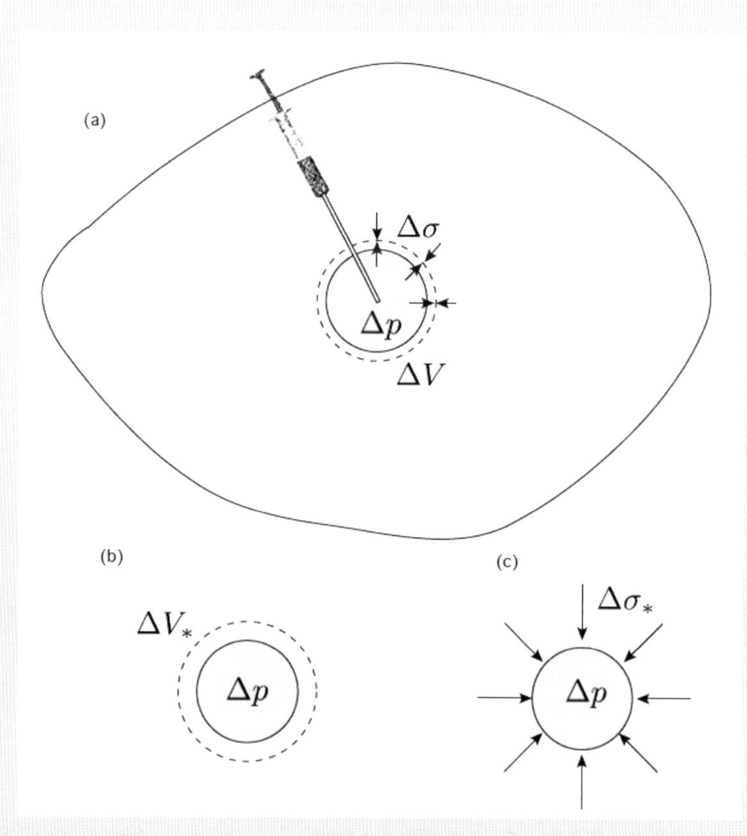

Figure (a) Injection of a finite volume of fluid inside the porous elastic sphere embedded in a large impermeable elastic body induces a pore pressure increase Δp inside the sphere as well as a stress perturbation $\Delta\sigma$ inside and outside the sphere, caused by the expansion ΔV of the sphere. (b) If the sphere is freed from its elastic surrounding, it will expand by the amount ΔV_* due to the pore pressure increase Δp. (c) A confining stress $\Delta\sigma_*$ needs to be applied on the free sphere to prevent the expansion ΔV_* caused by Δp. If the material in the surrounding medium is much softer than the material in the sphere, then $\Delta V \simeq \Delta V_*$ and $\Delta\sigma \simeq 0$; if the medium is much stiffer, then $\Delta V \simeq 0$ and $\Delta\sigma \simeq \Delta\sigma_*$. $\Delta\sigma$ refers only to the radial stress in the exterior region. Note: Syringe based on a concept from H.F. Wang (Wang, 2000).

cause a reduction of the slip resistance of a fault located in the region of pore pressure increase. In assessing the potential for induced seismicity, two basic questions arise: (1) What is the magnitude of the pore pressure change? and (2) What is the extent of the volume of rock where the pore pressure is modified in any significant manner? The magnitude of the induced pore pressure increase and the extent of the region of pore pressure change depend on the rate of fluid injection and total volume injected, as well as on two hydraulic properties of the rock, its intrinsic permeability (k) and its storage coefficient (S), and on the fluid viscosity (μ).

The permeability (k) is a quantitative measure of the ease of fluid flow through a rock; it depends strongly on the porosity of the rock (the volume percentage of voids in the rock volume) but also on the connectivity between pores. The storage coefficient (S) is a measure of the relative volume of fluid that needs to be injected in a porous rock in order to increase the pore pressure by a certain amount; the storage coefficient depends on the rock porosity in addition to the fluid and rock compressibility. The permeability (k) can vary by many orders of magnitude among rocks; for example, the permeability of a basement rock such as granite could be up to a billion times smaller than the permeability of oil reservoir sandstone (Figure 2.1).

However, the storage coefficient increases only by about one order of magnitude between a tight basement rock and high-porosity sandstone. The ratio $k{:}\mu S$ is the hydraulic diffusivity coefficient (c), which provides a measure of how fast a perturbation in the pore

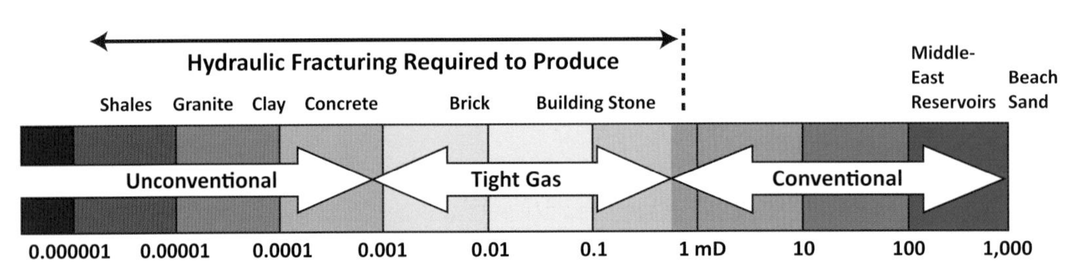

FIGURE 2.1 Comparison of permeability in oil and gas reservoirs utilizing permeability values for typical rock types and common building materials. The higher the connectivity between the pore spaces, the higher the permeability; for oil and gas reservoirs, higher permeability generally indicates greater ease with which the hydrocarbons will flow out of the reservoir and into a production well. Permeability is most commonly measured using a unit called a millidarcy (mD), and permeabilities can range between 1,000 mD (high permeability, comparable to beach sand) to very low permeability (0.000001 mD, which would describe the least permeable rocks such as shales). Other common materials (such as granite or brick) are noted on the upper part of the scale in this figure to give a sense of the range of permeabilities on the millidarcy scale. Although hydraulic fracturing has been used for decades to stimulate some conventional reservoirs, hydraulic fracturing is required to produce from low-permeability reservoirs such as tight sands and shales (left-hand side of the diagram). SOURCE: Adapted from King (2012).

pressure propagates in a saturated rock; like the permeability (k), the diffusivity (c) can vary over many orders of magnitude for different rocks. These parameters can be determined either from laboratory tests on drill core samples from wells or from pumping or injection tests, which have the advantage of providing estimates that are averaged over a scale relevant for reservoir calculations.

The intrinsic permeability of basement rocks is so low that the transport of fluid in these rocks can be thought of as taking place almost exclusively in the network of fractures that is pervading the crust. In other words, the rock itself can be viewed as being impermeable. Concepts of permeability and storage coefficient can be extended to fractures, where they transform into a transmissivity and storativity, with their ratio also having the meaning of diffusivity (see, e.g., Nicholson and Wesson, 1990; NRC, 1996).

The important point is that faults and fractures in basement rocks offer relatively little resistance to flow, and thus the equivalent permeability and diffusivity of these fractured rocks (with fractures and rocks viewed as a whole) can be very high. For example, the hydraulic diffusivity deduced from the time evolution of spatial spread of microseismic events measured during injection of water into a crystalline rock at Fenton Hill, an EGS site (Fehler et al., 1998), is about 0.17 m^2/s (Shapiro et al., 2003), a value in the range of those for very permeable sandstones. The combination of high transmissivity, small storativity, and the planar nature of fractures implies that significant pore pressure changes can be transmitted over considerable distances (several kilometers [miles]) through a fracture network from an injection well.

In permeable rocks, where the fluid is dominantly transported by a connected network of pores, the injection of fluid from a well can be viewed as giving rise to an expanding "bulb," centered on the well, which represents the region where the pore pressure has increased. The increase in pore pressure *decreases* with distance from the well until it becomes about equal to the initial pore pressure, prior to injection, at the edge of this expanding region. Once the size of this bulb becomes larger than the thickness of the permeable layer, the shape of this region becomes approximately cylindrical over the height of the layer. The region of perturbed pore pressure continues to grow radially until it meets bulbs growing from other injection wells or until it reaches the lateral boundaries of the reservoir (see also Nicholson and Wesson, 1990).

The dependence of the magnitude of induced pore pressure and of the size of the perturbed pore pressure region on the injection rate, the volume of fluid injected, and the rock hydraulic properties (permeability and storage coefficient) is complex. Numerical simulations are generally needed to establish these relationships, which depend on the geometry of the permeable rock. However, some general rules apply either at the early stage of injection when the bulb of increased pore pressure grows unimpeded by the interaction with the lateral boundaries of the reservoir or with other bulbs, or at a late stage of injection when the increase of the pore pressure is nearly uniform in the reservoir, which is here assumed to be of finite extent (see Appendix H for the calculation of the pore pressure induced by injection into a disc-shaped reservoir).

At the early stage of injection, the size of the bulb will essentially depend on the diffusivity of the rock and on the duration of injection (equal to the ratio of injected volume over the injection rate). The maximum induced pore pressure is equal to the ratio of the injection rate over the permeability times a function of the duration of injection. This means that the bulb size increases but the maximum pore pressure decreases with increasing rock permeability, everything else being equal. In other words, the induced pore pressure dissipates faster with increasing permeability. At the late stage of injection, the induced pore pressure does not depend on the injection rate and on the permeability, because it becomes proportional to the ratio of the volume of fluid injected over the storage coefficient.

The extent of the induced pore pressure field and the magnitude of the induced pressure are both relevant when assessing the risk of induced seismicity. A larger pore pressure increase brings the system closer to the conditions for initiating slip on a suitably oriented fault, if such a fault exists; a larger region of disturbed pore pressure will increase the risk of intersecting and activating a fault.

Inducing a significant seismic event requires an increase of the pore pressure above levels that have existed prior to fluid injection and over a region large enough to encompass a fault area consistent with the magnitude of the earthquake. For example, an earthquake of magnitude **M** 3 results from a rupture area of about 0.060 km^2 (corresponding to 15 acres). Such a situation was encountered at the Rangely, Colorado, oilfield starting in 1957, when sustained waterflooding operations (secondary recovery to improve petroleum production) over a period of several years caused the pore pressure to increase (Box 2.4). Eventually, pore pressure reached a level about 17 MPa (170 bars)[2] above the preproduction pore pressure, a threshold at which a series of seismic events began to occur; the largest of these events was **M** 3.4. However, waterflooding would not be expected to cause any significant seismic activity if the pore pressure did not exceed the initial pore pressure in a reservoir. Operators generally do not exceed preproduction pore pressure during waterflooding projects because they tend to maintain relative balance between the volumes of fluid injected and extracted. Exceptions to this generally balanced condition for waterflooding and resulting induced seismicity are cited in Appendix C.

Observations and monitoring of hydraulic fracturing treatments indicate that generally only microseismic events (microseisms, **M** < 2.0; see Chapter 1) are produced because the volume of fluid injected is relatively small (see also Chapter 3 for further details). Despite the fact that hydraulic fracturing does increase pore pressure above the minimum in situ stress (typically σ_h), the area affected by the increase in pore pressure is generally small, remaining in the near vicinity of the created fracture.

[2] MPa = megapascal; 1 MPa is equivalent to 10 bars or about 10 atmospheres of pressure.

SEISMICITY INDUCED BY FLUID WITHDRAWAL

Fluid extraction from a reservoir can cause declines in the pore pressure that can reach hundreds of bars. The declining pore pressure causes large contraction of the reservoir, which itself induces stress changes in the surrounding rock (Segall, 1989), in particular increasing horizontal stresses above and below the reservoir that could lead to reverse faulting (Figure 2.2). Grasso (1992) estimates that volume contraction of reservoirs from fluid withdrawal can cause earthquakes up to **M** 5.0.

Several examples of induced seismicity associated with fluid withdrawal and associated pore pressure decrease have been reported, notably at the Lacq gas field in France (Box 2.5). A study of induced seismicity associated with natural gas extraction in the Netherlands (Van Eijs et al., 2006) indicates that the three most important factors in producing seismicity are the pore pressure drop from pumping, the density of existing faults overlying the gas field, and the contrast in crustal stiffness between the reservoir rock and the surrounding rock.

Another proposed mechanism for initiating slip on preexisting faults is linked to the reduction of the vertical stress on the layers underlying the reservoir from which a large mass of hydrocarbons has been extracted (McGarr, 1991). In this mechanism, the buoyancy force of the Earth's lithosphere will cause an upward movement in the part of the crust that has been unloaded, thereby inducing slip on preexisting faults at depth.

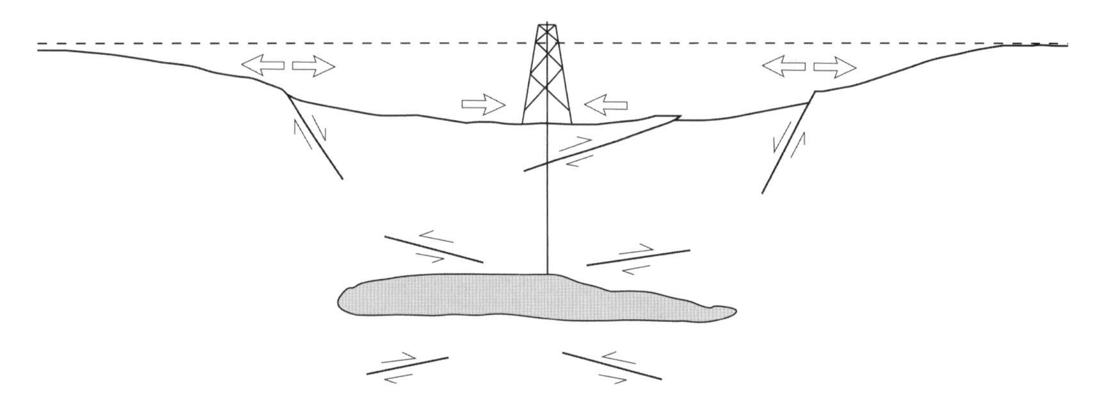

FIGURE 2.2 Observed faulting suggested to be associated with fluid withdrawal. Open arrows denote horizontal strain. In this interpretation, normal faults develop on the flanks of a field when the oil reservoir is located in a region of crustal extension. Reverse faults may develop above and below the reservoir if the reservoir is located in a region undergoing compression. Adapted after Segall (1989).

BOX 2.4
Induced Seismicity at the Rangely, Colorado, Oilfield

The Rangely, Colorado, induced seismicity experiment is an important milestone in the study of induced seismicity that firmly established the effective stress mechanism for induced seismicity. Water injection at the Rangely oilfield began in 1957 in response to declining petroleum production and decreased reservoir pressures. As a result of the waterflooding (secondary recovery) operations, reservoir pore pressures increased throughout the field, and by 1962 pore pressure in parts of the field substantially exceeded the original preproduction pressure of about 170 bars (17 MPa). In the same year the Uinta Basin Seismological Observatory, located about 65 km (~39 miles) from Rangely, began operation and detected numerous small seismic events **M** ≥ 0.5 in the vicinity of Rangely. With sustained fluid injection and elevated pore pressures the seismic events continued and the largest, **M** 3.4, occurred on August 5, 1964. Detailed monitoring with a local U.S. Geological Survey (USGS) seismic network installed in 1969 showed that the seismic events were occurring along a subsurface fault within the oilfield (Figure 1).

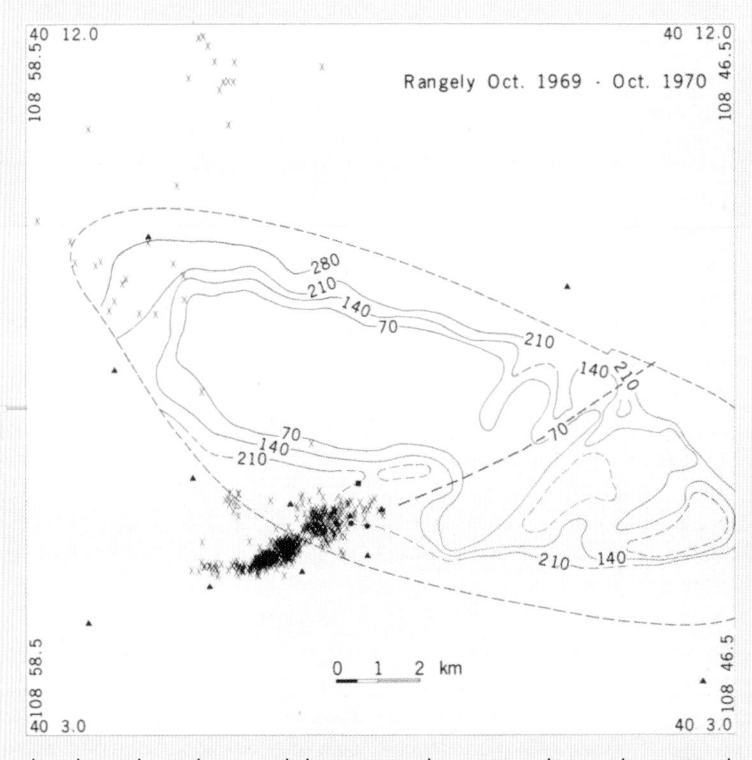

Figure 1 Earthquakes (x) located at Rangely between October 1969 and November 1970. The contours are bottom-hole 3-day shut-in pressures as of September 1969; the interval is 70 bars (7 MPa). Seismic stations are represented by triangles; experimental wells are represented by dots. The heavy, dashed line indicates the fault mapped in the subsurface. SOURCE: Raleigh et al. (1976).

With the cooperation of the Chevron Oil Company, which operated the field, USGS researchers carried out a controlled induced seismicity experiment beginning in November 1970 and continuing to May 1974 (Raleigh et al., 1976). One goal of the experiment was to quantitatively test the effective stress theory for activation of slip on preexisting faults by pore pressure increases (Box 2.1). This portion of the experiment entailed a program of careful measurements of the parameters involved in the Coulomb criterion (Box 2.1), including in situ stress measurements, monitoring and modeling of changes of reservoir pore pressures, laboratory measurement of the sliding resistance between rock surfaces in the reservoir formation where seismic events were occurring, and detailed seismic monitoring to precisely locate the events and determine the fault orientation with respect to the stress field. Together these measurements, when used with the Coulomb criterion expressed in terms of the effective stress, predicted that a critical reservoir pressure of 257 bars was required to induce earthquakes at an injection site within the cluster of earthquakes—a result that agreed with the observed and modeled pore pressures. The second phase of the experiment turned seismic events "on" and "off" by cycling the pore pressures above and below the critical reservoir pore pressure of 257 bars (25.7 MPa) (Figure 2). This experiment proved that induced seismic events could be controlled by regulating the pore pressures.

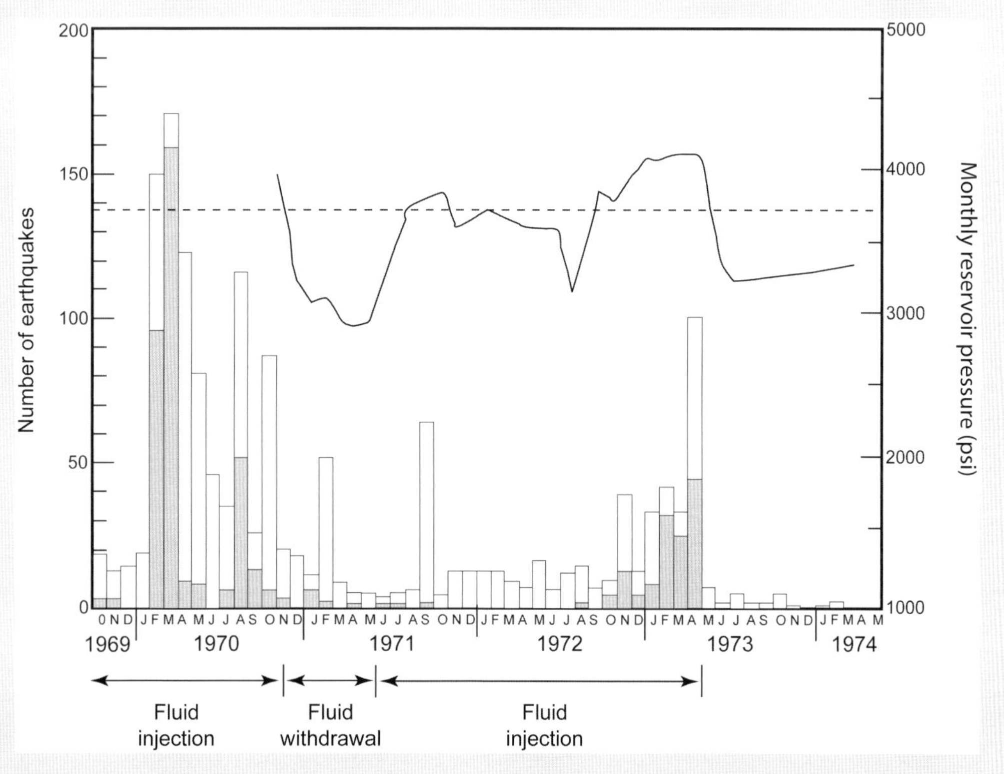

Figure 2 Frequency of seismic events at Rangely. Stippled bars are seismic events within 1 km of the experimental wells. The clear bars represent all other events. Pressure history in well Fee 69 is shown by the heavy line and predicted critical pressure is designated by the dashed line. SOURCE: Raleigh et al. (1976).

BOX 2.5
Induced Seismicity at the Lacq Gas Field (France)

The Lacq gas field in southwestern France offers one of the best-documented cases of seismicity induced by extraction of pore fluids (Grasso and Wittlinger, 1990; Segall et al., 1994). The gas reservoir is a 500-m-thick (~1,640-feet-thick) sequence of limestone that forms a dome-shaped structure at depths of 3.2 to 5.5 km (~2.0 to 3.4 miles) (Figure 1). The reservoir was highly overpressured when production started in 1957, with a pressure of about 66 MPa (660 bars) at a depth of 3.7 km (~2.3 miles) below sea level. The first felt earthquake took place in 1969, at a time when the pore pressure had decreased by about 30 MPa (300 bars). By 1983, the pressure had dropped by 500 bars, and 800 seismic events with magnitude up to **M** 4.2 had been recorded

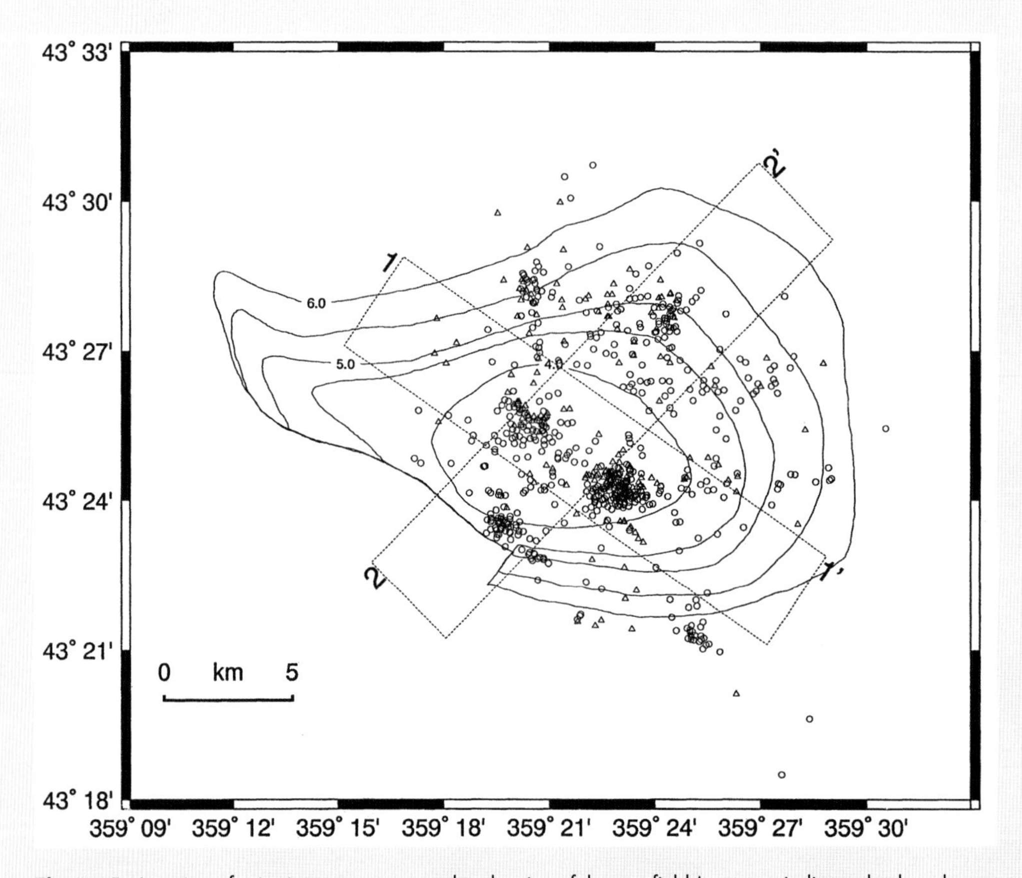

Figure 1 Location of seismic events compared to the size of the gas field (contours indicate depth to the top of the gas reservoir). Locations were determined from a local network and were based on an assumed velocity model. Triangles on the map are epicenters for events between 1976 and 1979; circles represent epicenters for events from 1982 to 1992. The rectangular areas (1-1′ and 2-2′) refer to other parts of the analysis conducted by Segall et al. (1994) and are not discussed further. SOURCE: After Segall et al. (1994).

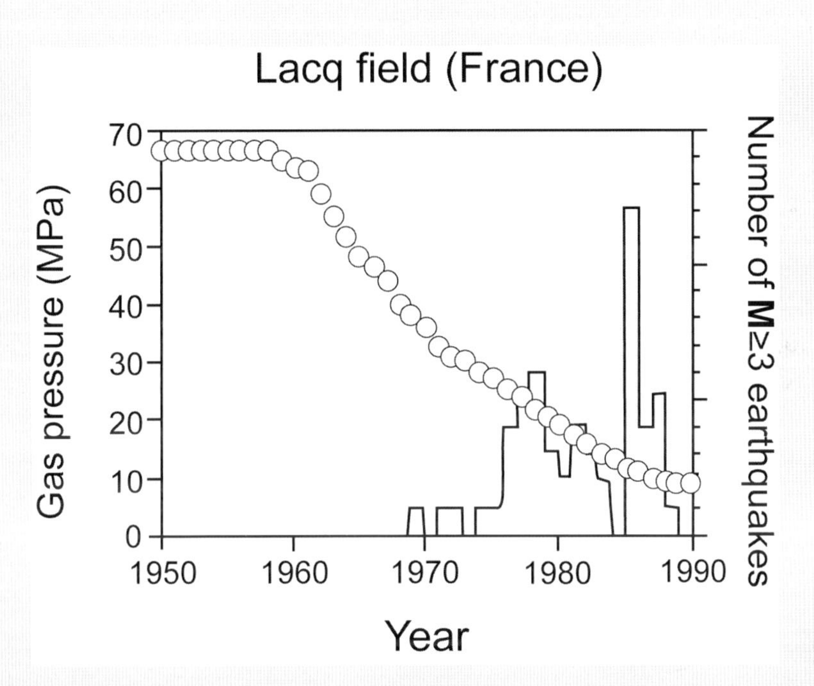

Figure 2 Decline of the pore pressure due to production at the Lacq gas reservoir and number of recorded earthquakes with magnitude **M** ≥ 3, with time, gas pressure (in MPa; 1 MPa is equal to 10 bars) (circles, left scale), and number of **M** > 3 earthquakes per year (solid line, right scale). The number of earthquakes increased with decreasing pressure. SOURCE: Segall (1989).

(Figure 2). The epicenters of 95 percent of the well-located events and all of the **M** > 3 events were within the boundaries of the gas field (Grasso and Wittlinger, 1990).

An analysis of the stress changes above and below the reservoir indicates that the induced seismicity is consistent with a thrust fault regime where the least compressive stress is vertical. Furthermore, the maximum shear stress change is calculated to be about 0.1 MPa (1 bar) for a pressure drop of 30 MPa (300 bars), suggesting that the in situ stress prior to production was close to causing frictional failure of the rock.

SOURCES: Segall (1989); Segall et al. (1994); Segall and Fitzgerald (1998); Grasso and Wittlinger (1990); Grasso (1992).

SUMMARY

Both the conditions that lead to the initiation of a seismic event and the factors that affect the magnitude of the resulting event are well understood. The conditions of initiation are embodied in the Coulomb criterion (involving a comparison of the shear stress on the fault to the fault frictional strength), while the magnitude of the seismic event is related to the area of the fault undergoing slip. Inducing a seismic event requires a triggering event that will either increase the shear stress or reduce the normal effective stress on the fault and/or reduce the fault frictional resistance, for example, an increase of the pore pressure that reduces the frictional strength to a level at which it is overcome by the driving shear stress. However, to cause a significant event requires activating slip over a large enough area; for example, a seismic event of **M** 4 involves a fault area of about 1.4 km^2 (~0.5 square miles) and a slip of about 1 m (~39 inches).

Unfortunately, despite our understanding of the factors affecting the initiation and the magnitude of a seismic event, the values of the process parameters (such as the injection rate or the volume of fluids injected) that will trigger the seismic event and what magnitude the event will be are generally not possible to quantify. The inability to make these kinds of predictions is due to several factors: (1) fragmentary knowledge of the state of stress in the Earth; (2) lack of knowledge about the faults themselves, including their existence (if they have not yet been mapped) and their orientations and physical properties; and (3) difficulty in collecting the basic data (hydraulic and mechanical parameters, geometry of the geological structure, such as the reservoir) that are required to calculate the pore pressure and stress change induced by the fluid injection or withdrawal.

Nonetheless, the insights into the mechanisms causing seismic events allow us to make some broad conclusions. In processes involving fluid injection, the pore pressure increase is the dominant factor to be considered, as stress change can often be ignored. Any increase of the pore pressure above historical undisturbed values may bring the system closer to critical conditions. The probability of triggering a significant seismic event increases with the volume of fluid injected: the larger the volume injected, the more likely a larger fault will be intersected. However, injection of fluid in depleted reservoirs (such as in secondary recovery stimulation—waterflooding) is unlikely to create an earthquake, irrespective of the volume of fluid injected, if the pore pressure remains below preproduction values.

The transient region of high pore pressure that surrounds a newly created hydraulic fracture is not expected to be large enough for a significant seismic event to be triggered, except in rare cases where the new hydraulic fracture intersects or is very near an existing fault. Even in such cases, the magnitude of the event is expected to be small because a large fault area will not be affected.

The fluid injected in crystalline basement rocks is essentially transmitted by a network of interconnected fractures and joints. Because of the high transmissivity and low storativity of these kinds of rocks, the potential exists to induce pore pressure increase at considerable distances from the injection well and thus trigger slip on faults that are located kilometers away from the injection source.

Seismicity induced by fluid withdrawal cannot be explained without taking into account the accompanying stress changes, which are associated with the large-scale contraction of the reservoir caused by pore pressure reduction or uplift caused by removal of a significant mass of hydrocarbons. The magnitude of the events can be potentially large, because the stress change takes place over areas that are similar in size to the reservoir. However, to trigger an earthquake requires the initial state of stress to be very close to critical, because the perturbation of the stress is minute compared to the magnitude of the pore pressure reduction. For example, in the well-documented Lacq gas field (France) the increase of the maximum shear stress was estimated to be about 0.1 MPa (1 bar) in regions surrounding the reservoir for a pressure drop of 30 MPa (300 bar) in the reservoir.

REFERENCES

Brown, E.T., and E. Hoek. 1978. Trends in relationships between in situ stresses and depth. *International Journal on Rock Mechanics* 15:211-215.

Fehler, M., L. House, W.S. Phillips, and R. Potter. 1998. A method to allow temporal variation of velocity in travel-time tomography using microearthquakes induced during hydraulic fracturing. *Tectonophysics* 289:189-202.

Grasso, J.-R. 1992. Mechanics of seismic instabilities induced by the recovery of hydrocarbons. *Pure and Applied Geophysics* 139(3-4):507-534.

Grasso, J.-R., and Wittlinger. 1990. 10 years of seismic monitoring over a gas field area. *Bulletin of the Seismological Society of America* 80:450-473.

Hanks, T.C., and H. Kanamori. 1979. A moment magnitude scale. *Journal of Geophysical Research* 84:2348-2350.

Heidbach, O., M. Tingay, A. Barth, J. Reinecker, D. Kurfeß, and B. Müller. 2008. *The World Stress Map database release 2008*, doi:10.1594/GFZ.WSM.

Hsieh, P.A. 1996. Deformation-induced changes in hydraulic head during ground-water withdrawal. *Ground Water* 34(6):1082-1089.

Jaeger, J.C., N.G.W. Cook, and R.W. Zimmerman. 2007. *Fundamentals of Rock Mechanics*, 4th ed. New York: Blackwell Publishing.

King, G.E. 2012. Hydraulic Fracturing 101: What every representative, environmentalist, regulator, reporter, investor, university researcher, neighbor, and engineer should know about estimating frac risk and improving frac performance in unconventional gas and oil wells. Paper SPE 152596 presented to the Society of Petroleum Engineers (SPE) Hydraulic Fracturing Technology Conference, The Woodlands, TX, February 6-8.

McGarr, A. 1991. On a possible connection between three major earthquakes in California and oil production. *Bulletin of the Seismological Society of America* 81(3):948-970.

NRC (National Research Council). 1990. *The Role of Fluids in Crustal Processes*. Washington, DC: National Academy Press.

NRC. 1996. *Rock Fractures and Fluid Flow: Contemporary Understanding and Applications*. Washington, DC: National Academy Press.

Nicholson, C., and R.L. Wesson. 1990. Earthquake hazard associated with deep well injection: A report to the U.S. Environmental Protection Agency. U.S. Geological Survey (USGS) Bulletin 1951. Reston, VA: USGS. 74 pp.

Raleigh, C. B., J.H. Healy, and J.D. Bredehoeft. 1976. An experiment in earthquake control at Rangely, Colorado. *Science* 191:1230-1237.

Scholz, C.H. 2002. *The Mechanics of Earthquakes and Faulting*. Cambridge: Cambridge University Press.

Segall, P. 1989. Earthquakes triggered by fluid extraction. *Geology* 17:942-946.

Segall, P., and S.D. Fitzgerald. 1998. A note on induced stress changes in hydrocarbon and geothermal reservoirs. *Tectonophysics* 289:117-128.

Segall, P., J.-R. Grasso, and A. Mossop. 1994. Poroelastic stressing and induced seismicity near the Lacq gas field, southwestern France. *Journal of Geophysical Research* 99(B8):15,423-15,438.

Shapiro, S.A., R. Patzig, E. Rothert, and J. Rindschwentner. 2003. Triggering of seismicity by pore-pressure perturbations: Permeability-related signatures of the phenomenon. *Pure and Applied Geophysics* 160(5):1051-1066.

Van Eijs, R.M.H.E., F.M.M. Mulders, M. Nepveu, C.J. Kenter, and B.C. Scheffers. 2006. Correlation between hydrocarbon reservoir properties and induced seismicity in the Netherlands. *Engineering Geology* 84:99-111.

Wang, H.F. 2000. *Theory of Linear Poroelasticity with Applications to Geomechanics and Hydrogeology*. Princeton, NJ: Princeton University Press.

Wells, D.F., and K.J. Coppersmith. 1994. New empirical relationships among magnitude, rupture length, rupture width, rupture area, and fault displacement. *Bulletin of the Seismological Society of America* 84(4):974-1002.

Zoback, M.L., and M.D. Zoback. 1980. State of stress in the conterminous United States. *Journal of Geophysical Research* 85:6113-6156.

Zoback, M.L., and M.D. Zoback. 1989. Tectonic stress field of the conterminous United States. *Memoirs of the Geological Society of America* 172:523-539.

CHAPTER THREE

Energy Technologies: How They Work and Their Induced Seismicity Potential

Much of the energy used in the United States comes from fluids pumped out of the ground. Oil and gas have been major energy sources in the country for over 100 years, and new developments in the production of natural gas indicate that it may provide a significant source of energy for the nation during the twenty-first century. Geothermal power has been used to supply energy in the United States for almost as long as oil, although major electricity generation from geothermal energy sources began only in the 1960s at The Geysers in Northern California. A 2006 report on the potential of geothermal energy (MIT, 2006) suggested it could be a major contributor to the nation's energy supply in the coming decades. Efforts to reduce concentrations of carbon dioxide (CO_2) in the atmosphere have spurred development of technologies to capture and store (sequester) CO_2. Projects to accomplish carbon capture and storage (CCS) from industrial facilities are currently being piloted in the United States and elsewhere in the world. Underground injection of CO_2 has also been commonly used to enhance oil and gas recovery.

This chapter reviews the potential for induced seismicity related to geothermal energy production, conventional oil and gas development (including enhanced oil recovery [EOR]), shale gas development, injection wells related to disposal of wastewater associated with energy extraction, and CCS.

GEOTHERMAL ENERGY

Geothermal energy exists because of the substantial heat in the Earth and the temperature increase with depths below the Earth's surface. Depending upon the regional geology—including the composition of the rocks in the subsurface and any of the fluids contained in the rocks—the temperature increase with depth (the thermal gradient) may be fairly steep and represent the source of sufficient geothermal energy to allow commercial development for electricity generation. The largest actively producing geothermal field in the United States at The Geysers in Northern California generates approximately 725 megawatts of electricity per year ("megawatts electrical" or MWe). This is enough to power 725,000 homes or a city the size of San Francisco. Currently this geothermal field supplies nearly 60 percent of the average electricity demand of the northern coastal region of California.

The most likely regions for commercial development of geothermal power are generally the same regions that have experienced recent volcanism (Figure 3.1). Such areas are concentrated in the western portion of the country. The U.S. Geological Survey (USGS) estimates that the total power output from the hydrothermal (vapor- and liquid-dominated) geothermal resources in the United States can probably be increased to 3,700 MWe per year, and a 50 percent probability exists that it can be increased to about 9,000 MWe per year (Williams et al., 2008). Two recent studies have produced nationwide estimates of the electric power potential that might be achieved by a successful implementation of enhanced geothermal systems (EGS) technology, perhaps contributing 100,000 MWe of electrical power per year (MIT, 2006). More recently the USGS (Williams et al., 2008) has published a mean estimate for potential EGS development on private and accessible public land at

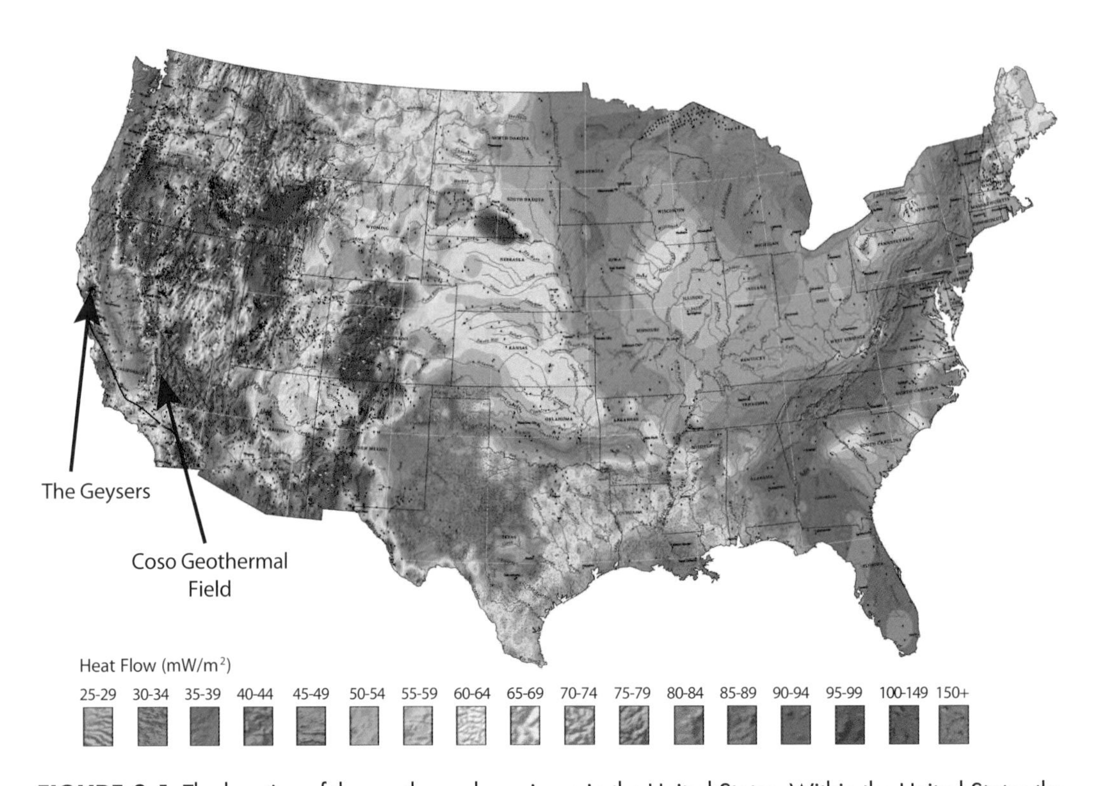

FIGURE 3.1 The location of the geothermal provinces in the United States. Within the United States the regions of relatively high thermal gradients, shown in red, exist only in the West. The typical local geologic setting for these high-geothermal-gradient areas is within sedimentary basins located near or intruded by recent volcanics, or within (as part of) the buried volcanic rocks themselves. Only one vapor-dominated reservoir has been developed in the United States (The Geysers); the remainder of the areas in red and orange may host viable liquid-dominated or enhanced geothermal system reservoirs. SOURCE: SMU Geothermal Lab; Blackwell and Richards (2004).

517,800 MWe. This is approximately half of the current installed electric power generating capacity in the United States.[1]

The three different forms of geothermal resources are recognized: (1) "vapor-dominated," where primarily steam is contained in the pores or fractures of hot rock; (2) "liquid-dominated," where primarily hot water is contained in the rock; and (3) "hot dry rock," where the resource is simply hot and currently dry rock that requires an EGS to facilitate development (see Figure 2.1). Vapor- and liquid-dominated systems are collectively termed hydrothermal resources. The vast majority of known hydrothermal resources are liquid dominated.

The different forms of geothermal resources result in significant differences in the manner in which they are developed and particularly in the manner that liquids are injected to help stimulate energy development. Different injection practices can cause induced seismicity through different processes. The nature of and differences among the induced seismicity that may result from each of the three geothermal resources are summarized here.

Vapor-Dominated Geothermal Resources

A limited number of localities in the world exist where the geothermal resources naturally occur as steam. Despite their rarity, the two largest geothermal developments of any kind in the world are both vapor-dominated geothermal reservoirs. The Larderello geothermal field in the Apennine Mountains of northern Italy became the first of these and has generated electricity continuously since 1904, except during World War II. However, the most productive geothermal field development in the world is The Geysers (Figure 3.2), located about 75 miles north of San Francisco. The Geysers also has the most historically continuous and well-documented record of seismic activity associated with any energy technology development in the world.

The first commercial power plant at The Geysers came online in 1960 with a capacity of 12 MW (Koenig, 1992). Over the next 29 years the installed generation capacity was increased to a total of 2,043 MW through building 28 additional power plant turbine-generating units (CDOGGR, 2011). The basic elements of the process to generate electricity in this type of power plant are illustrated in Figure 3.3.

These plants were supplied with steam from 420 production wells, with the steam capable of flowing up the production wells under its own pressure. The condensed steam not evaporated at the power plant cooling towers was being reinjected into the steam reservoir by using 20 injection wells drilled to similar depths. The area of development had been expanded from the original 3 square miles to about 30 square miles. Because the generation of energy from the field consumes natural steam originally in the reservoir, by 1988

[1] See http://www.eia.gov/electricity/capacity.

FIGURE 3.2 Ridgeline Unit 7 and 8 Power Plant (rated at 69 MW) in the left foreground at The Geysers in California. The turbine building, housing the two turbine-generator sets, the operator's control room, and various plant auxiliaries are on the left. The evaporative cooling tower with steam emanating from the top is on the right of the main complex. The beige pipelines along the roads (with square expansion loops) are the steam pipelines that gather the steam from the production pads and bring it to the plant. A high-voltage transmission line (denoted by lattice towers) is in the middle foreground of the picture. SOURCE: Calpine.

the production of steam had started to decline; this decline was marked by a significant decrease in reservoir pressure from an original pressure of about 500 pounds per square inch (psi)[2] to levels as low as 175 psi (Barker et al., 1992). For years the annual injection volumes returned to the geothermal reservoir were less than a third of the amount of steam being produced, so the reservoir was drying up. New sources of water were established by constructing two pipelines that currently deliver about 25 million gallons of treated wastewater a day for injection, increasing the current annual mass replacement to 86 percent compared to 26 percent back in 1988 (CDOGGR, 2011).

Early reports of induced seismicity at The Geysers, begun by USGS researchers (Hamilton and Muffler, 1972), described microseismicity that was observed close to where

[2] A car tire for a standard, midsized automobile is usually inflated to a pressure of about 30-35 psi for comparison.

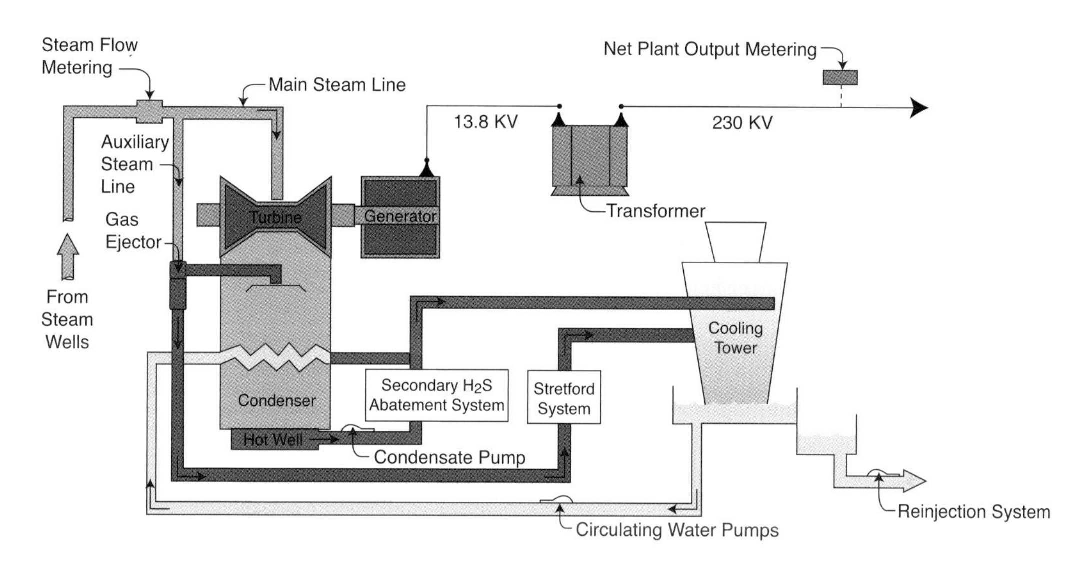

FIGURE 3.3 Elements of the power plant cycle for vapor-dominated geothermal resources. The steam is directed by the main steam line into a turbine that spins the connected generator unit, typically generating electricity at 13.8 kilovolts (kV), which a transformer increases to 230 kV for distribution by a transmission line. The steam leaving the turbine enters the condenser that contains a network of tubing through which cool water is circulated, facilitating the condensation process. The condensate is then pumped to the cooling tower where it is cooled by evaporation, with the cooled water being in part recirculated by the circulating water pumps back to and through the condenser. Because some noncondensable gases usually occur naturally in the steam, those gases are removed from the condenser by the gas ejector system that creates a partial vacuum by the flow of a small amount of steam delivered by the auxiliary steam line. Those gases, in particular H_2S, are chemically processed commonly by a Stretford System before delivery to the cooling tower where they are vented. SOURCE: Adapted from the Northern California Power Agency.

the geothermal development operations were taking place. As the area of steam field development expanded, the areal distribution of seismic events similarly expanded, and the number of the events progressively increased (Figure 3.4).

With the addition of more seismometers of increased sensitivity distributed throughout the expanded development area, a clear association became evident between these induced events and the active injection wells and volume of water being injected. Figure 3.5 shows where injection took place in the southeastern part of The Geysers in 1998, the year following the startup of the first wastewater pipeline that more than doubled the injection volume. During 1997-1998, 1,599 events of $\mathbf{M} \geq 0.6$ were recorded, an increase of just over 50 percent compared to the prior 12 months.

The history of steam production, water injection, and seismic history at The Geysers since 1965 is shown in Box 3.1. Steam production and therefore electricity generation reached a maximum in 1987, followed by a fairly rapid decline until the wastewater pipelines

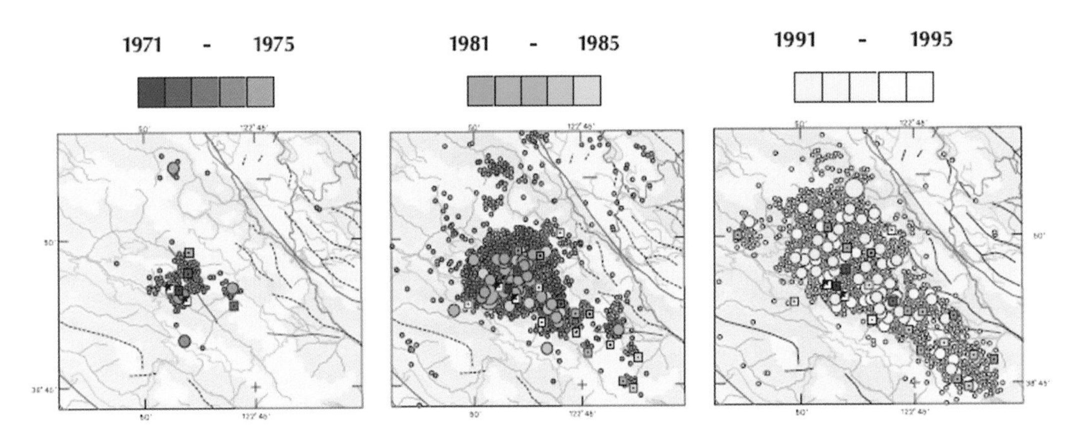

FIGURE 3.4 Geysers seismicity maps in 10-year intervals show the expanding distribution of development as illustrated by the increased numbers of green squares that indicate the locations of the operating power plants. SOURCE: Preiss et al. (1996).

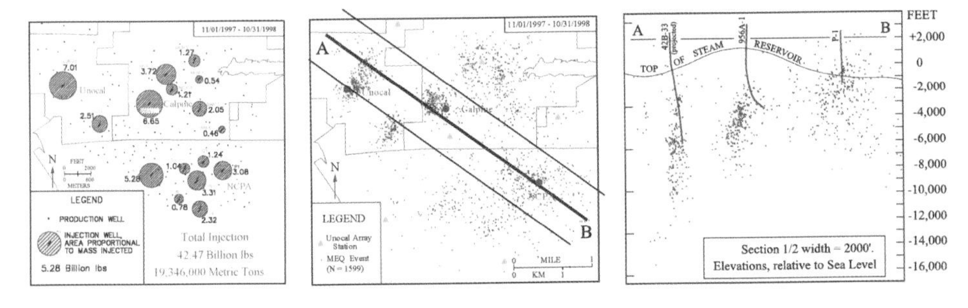

FIGURE 3.5 The locations of injection wells and the location and depth distribution of seismic events in the southeastern part of The Geysers area during 1997-1998. Map on the left shows injection wells in 1998. The middle map shows the total number of recorded seismic events from the period 1997-1998 with the line of cross section (figure on the right). The cross section shows the positions of three geothermal wells with the location at depth of the seismic events (red dots). SOURCE: Beall et al. (1999).

began deliveries in 1997 and 2003. The annual amount of water injected followed the same trends until new sources of water other than condensate were developed, allowing recent injection to become nearly equal to the annual production levels.

The method of injection at The Geysers is unusual because of the extremely low fluid pressures in the deep underlying reservoir. No surface pressure is needed to inject; the water simply falls down the injection well as though through a partial vacuum because the fluid pressures in the reservoir are incapable of supporting a liquid level to the surface. Consequently, without elevated bottom-hole pressures, the primary cause of the induced

BOX 3.1
Geysers Annual Steam Production, Water Injection, and Observed Seismicity, 1965-2010

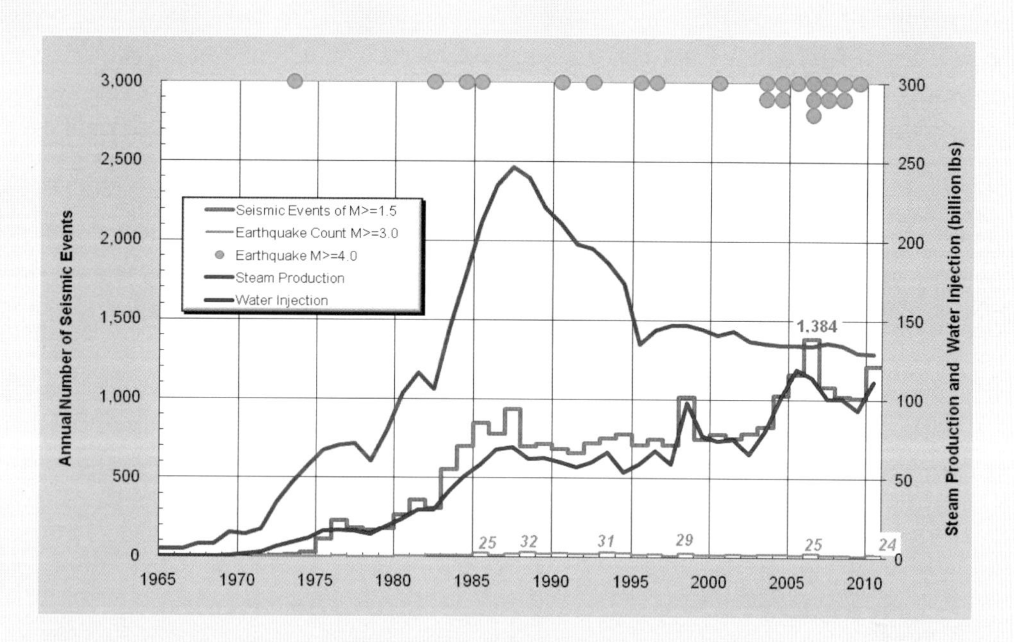

Figure The history of induced seismicity at The Geysers is shown in three forms. First, the number of recorded events of **M** 1.5* and greater is shown to have increased from almost none in the 1960s to 112 in 1975 and then to as many as 1,384 in 2006 (thick green line). Second, the annual number of earthquakes of **M** 3.0 and greater is shown along the bottom of the graph (pale green line). By 1985, 25 such events occurred annually, and that rate of about two events of **M** 3.0 and greater per month has continued to the present. Third, events of **M** 4.0 and greater are shown near the top (green dots). The first such event occurred in 1972, and more recently about one to three of these have occurred per year. The maximum magnitude was a **M** 4.67 event in May 2006. SOURCES**:** Adapted from Smith et al. (2000) and Majer et al. (2007).

*Note that this report uses **M** 2.0 as the general limit below which earthquakes cannot be felt by humans; however, at The Geysers **M** 1.5 is the lowest magnitude that the USGS can report faithfully year after year. Furthermore, residents in Anderson Springs may feel events as low as **M** 1.5 because the events are spatially quite close to the community.

seismicity is the fact that the hot subsurface rocks are significantly cooled by the injected water, and the resulting thermal contraction reduces the confining pressures and allows the local stresses to be released by limited movement on fracture surfaces.

The two strong motion recording instruments installed in 2003 near the neighboring communities of Anderson Springs and Cobb commonly record moderate shaking, plus about a dozen Mercalli VI (strong shaking) events each year (see also Chapter 1 for a definition of the Mercalli scale). The one event of Mercalli VII intensity caused an average acceleration of 21.0%g[3] at Anderson Springs and was related to a **M** 3.03 seismic event located at a depth of 4,750 feet only 1.2 miles west of the recording instrument.

The operators at The Geysers meet regularly with representatives of these two communities, county government, federal and state regulatory agencies, the USGS, and the Lawrence Berkeley National Laboratory to discuss the field operations and the recently observed seismicity. Minor damage is occasionally caused by the induced seismicity at The Geysers, generally as cracks to windows, drywalls, or tile walls or flooring in these communities. A system for receiving, reviewing, and approving such damage claims attributed to the local seismicity was established 6 years ago, and the homeowners are reimbursed for their costs to have the home damage repaired. To date these reimbursements for home repairs total $81,000, and this system appears to be resulting in mutually satisfactory relationships.

Liquid-Dominated Geothermal Resources

In contrast to the development of the vapor-dominated geothermal resources, liquid-dominated resources commonly use downhole pumps in the production wells to deliver the thermal waters to surface facilities. Surface pumping facilities are needed to force the injected waters back down into the reservoir. The liquid-dominated geothermal reservoirs that have been commercially developed to produce electricity in the western United States are listed in Table 3.1 (sources include the California Division of Oil, Gas and Geothermal Resources [CDOGGR], the Nevada Commission on Mineral Resources, the Imperial Irrigation District, and various operators).

Several different methods are used to generate electricity in liquid-dominated geothermal systems depending primarily on the temperature of the produced fluids; the flash steam power cycle process and the binary cycle process are the most common (Figure 3.6).

The cause and extent of the induced seismicity related to the development of liquid-dominated geothermal resources are different from those in the vapor-dominated resources (Box 3.2). From the start of operations the amount of fluid produced from a liquid-dominated reservoir is almost fully replaced by injection, which prevents a signifi-

[3] "%g" is motion measured as acceleration by an instrument, expressed as a percent of the acceleration of a falling object due to gravity.

cant decline in reservoir pressure. The temperature difference between the produced and reinjected waters is also relatively limited, so less cooling of the reservoir results. Consequently, if the surface and resulting bottom-hole pressures in the injection wells are limited to be less than that necessary to induce fracturing, little cause exists for the operations to produce significant induced seismicity. Monitoring at many of the liquid-dominated geothermal fields has demonstrated a relative lack of induced seismicity. However, as described below, the Coso geothermal field began as a strictly liquid-dominated field and has evolved during extended production to become partly vapor dominated. This evolution has resulted in reduction in fluid replacement and has caused the introduction of induced seismic events.

The Coso geothermal field provides a well-documented example of a complex resource area that was liquid dominated before the start of development 25 years ago and that may have evolved, following extensive production, into a resource that is now in part vapor dominated (see Box 3.2). Coso near Ridgecrest, in southeast-central California, is in a region of recent volcanism that is also seismically active. The first commercial geothermal power plant began operating in 1987; since 1989 three plants have been in operation with a total generating capacity of 260 MW, with about 85 production and 20 injection wells currently in use (CDOGGR, 2011). The geothermal fluids (dominantly water) are at temperatures in excess of 300°C (572°F) at depths of 1.5-2 km (~0.9-1.2 miles) (Feng and Lees, 1998).

The areal coincidence of the local seismicity at Coso with local surface subsidence, identified by using synthetic aperture radar data, suggest that the Coso field operations have caused reservoir cooling and thermal contraction, resulting in induced seismicity (Fialko and Simons, 2000). More recently, Kaven et al. (2011), based in part on their investigation of local changes in seismic velocities (V_p:V_s ratios), attribute the induced seismicity at Coso to decreases in fluid saturation and/or fluid pressure within the active geothermal reservoir.

An important issue to emphasize with regard to potential changes in pore pressure at vapor- and liquid-dominated geothermal power plants is the selection of conversion cycle—whether flash cycle or binary cycle (see Figures 3.2 and 3.6). The cycle selection is determined by the temperature and nature (physical state) of the geothermal fluids produced to the surface. Those power-cycle differences are important to explain why evaporative losses are significant at vapor-dominated resource power plants and moderate at flash cycle power plants. Evaporative losses can result in pore pressure and thermal losses that in turn can result in significant or moderate levels of induced seismicity. Equally important is to explain why in the case of binary cycle power plants there are no evaporative losses and generally little if any loss of pore pressure or fluid temperature, and therefore little if any associated induced seismicity.

TABLE 3.1 Liquid-Dominated Geothermal Fields in the United States with Operating Power Plants

Area Field	Plant Start Year	Power Cycle Used	Power Plant Capacity (MWe)	Average Generation (MWe)	Average Resource Temperature (°F)	Owner/Operator
California						
Imperial Valley						
North Brawley	2010	Binary	50	20.9	375	Ormat
East Mesa	1987	Binary & Flash	105	59	306	Ormat
Heber	1985	Binary & Dual Flash	92	75.9	324 to 350	Ormat
Salton Sea	1982	Single, Dual, & Triple Flash	352	314.6	480 to 690	Cal Energy
Mojave Desert						
Coso	1987	Dual Flash	260	48	480 to 580	TerraGen
Mammoth						
Casa Diablo	1984	Binary	29	20.9	340	Ormat
Power subtotal			888	539.3		
Nevada						
Reno/Fallon						
Brady	1992	Dual Flash	26.1	14.8	284	Ormat
Desert Peak	2006	Binary	14	14	370	Ormat
Jersey Valley	2010	Binary	15	Na	330	Ormat
Salt Wells	2009	Binary	28	Na	na	Enel
San Emidio	1987	Binary	3.6	2.6	275 to 290	U.S. Geothermal
Soda Lake	1987	Binary	26.1	10.4	360 to 390	Magma
Steamboat	1988	Binary & Flash	139.5	105.5	300	Ormat
Stillwater	2009	Binary	47.3	15.9	na	Enel
Wabuska	1987	Binary	2.4	0.8	na	H.S. Geothermal

North Central						
Beowawe	1985	Dual flash	16.6	14.6	410	TerraGen
Blue Mountain	2009	Binary	49.5	40	375	Nevada Geo
Dixie Valley	1988	Dual flash	67.2	41.2	400 to 480	TerraGen
Power subtotal			435.3	259.8		
Utah						
Roosevelt	1984	Binary & Flash	37	34	510	Pacific Corp
Thermo	2008	Binary	10.0	6.6	250 to 390	Raser
Power subtotal			47	40.6		
Idaho						
Raft River	2008	Binary	13	8.4	275 to 300	U.S. Geothermal
Hawai'i						
Big Island						
Puna	1993	Combined Cycle	30	na	330	Ormat
Alaska						
Fairbanks area						
Chena Hot Springs	2006	Binary	0.73	0.5	165	Chena Energy
Power totals			1414.03	848.6		

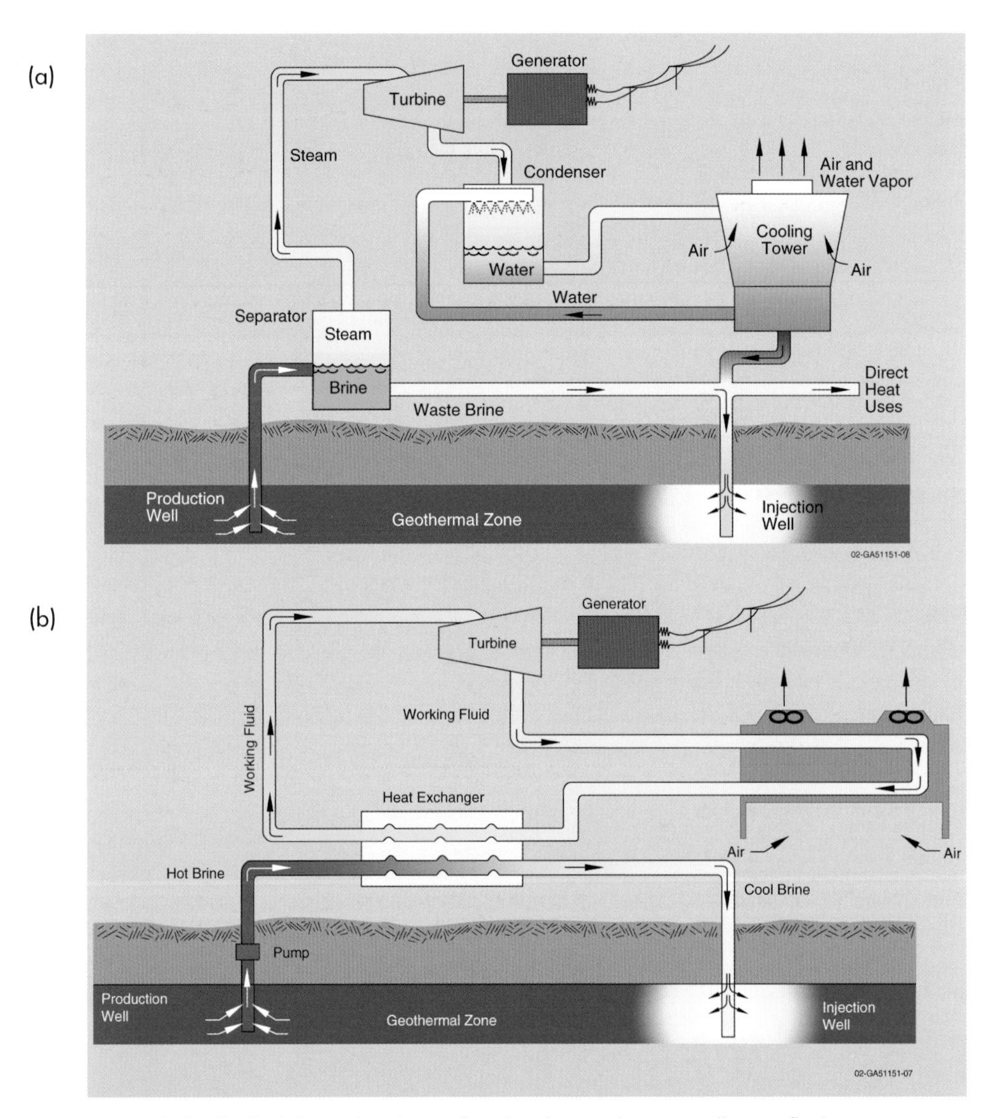

FIGURE 3.6 (a) The fluids delivered to the surface by the production wells in a flash steam power cycle are passed through a flash vessel or separator; the separated steam that flows out of the top is directed into a power plant where it is used to spin a steam turbine connected to a generator that produces an electrical output. The spent steam travels through a condenser, and the condensate is then pumped to the cooling tower, where the liquids are cooled before some of the fluids are pumped back inside the condenser and some are combined with the water drained from the bottom of the separator and sent to the injection wells. (b) The produced fluids for binary cycle power plants are first passed through a heat exchanger to heat a secondary liquid, usually an organic fluid such as isopentane, which vaporizes (boils) at a lower temperature than does water. That vaporized secondary fluid is then used to spin a turbine generator to make electricity. Similarly, that vapor is then condensed and returned directly to the heat exchanger to be reheated, revaporized, and recycled without any fluid loss. The produced geothermal water that has passed through the heat exchanger is then delivered to the injection wells. SOURCE: Idaho National Laboratory.

BOX 3.2
Induced Seismicity at the Coso Liquid-Dominated Geothermal Field

Locally induced seismicity recorded in the area of the Coso geothermal field development between 1996 and 2008 in map view (Figure 1, top) and cross section (Figure 1, bottom) shows clustering relative to the location and depth of the geothermal wells shown in blue. The number of seismic events of magnitude 0.5 and greater is plotted; these events total 10,200.

The history of geothermal fluid (dominantly water) production, water injection, and recent seismic history at the Coso field from 1977 through 2011 is shown in Figure 2. Starting in 1987, annual production reached a maximum of 121 billion lb* in 1990 and had decreased to 68 billion lb by 2009, while annual injection has declined from a maximum of 80 billion lb to 27 billion lb (CDOGGR, 2011). The relatively low reinjection rate for a liquid-dominated resource is because of cooling tower evaporative losses that result from the produced fluids containing an increased steam fraction as reservoir pressures have declined over the almost 25 years of operation.

Using the catalog of data available from the Southern California Earthquake Data Center, the history of local seismicity at the Coso field from 1977 to 2009 is shown in Figure 2.

With reference to Figure 2, the number of events of **M** 1.5 and greater averaged 5 per year during the 10 years prior to development, then doubled in the first 5 years after 1987, reaching maxima of 51 in 1995, 55 in 1998-1999, and 64 in 2001 before declining to a current level of about 20 per year. The peaks in 1995 and in 1998-1999 were attributed by Bhattacharyya and Lees (2002) to triggering in response to significant (**M** > 5.0) nearby earthquakes at Ridgecrest and in the Coso range. Additionally, the number of earthquakes of **M** 3.0 and greater is shown near the bottom of the chart. Single events occurred in 1978, 1995, 1998, 1999, and 2007, with three in 2009. The single earthquake in 2007 was a **M** 4.11 event, as shown near the top of the chart.

---

*Note that where at least part of the production is in the form of steam as well as liquid water, "pounds" is needed as the single unit to describe both the quantity of production and injection because gallons or cubic meters cannot be used in reference to steam.

(Box continues)

BOX 3.2 Continued

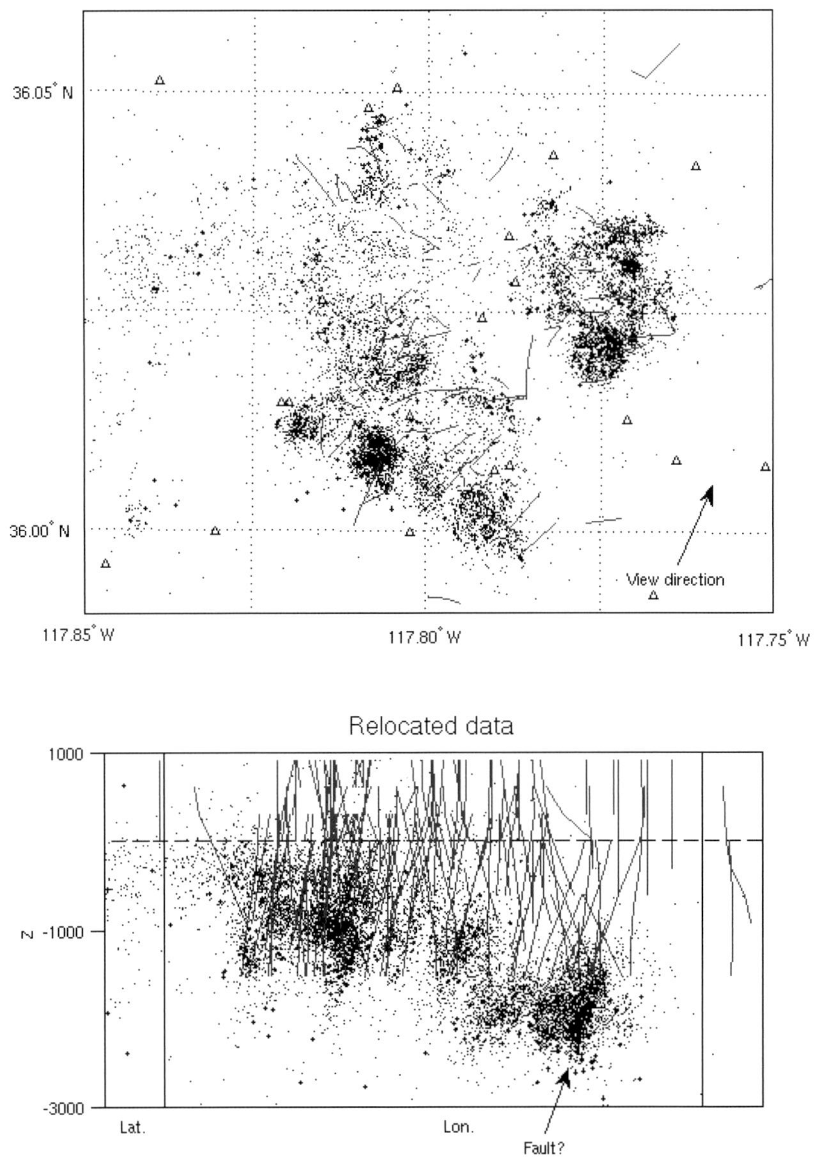

Figure 1 Seismicity recorded at the Coso geothermal field. SOURCE: Kaven et al. (2011).

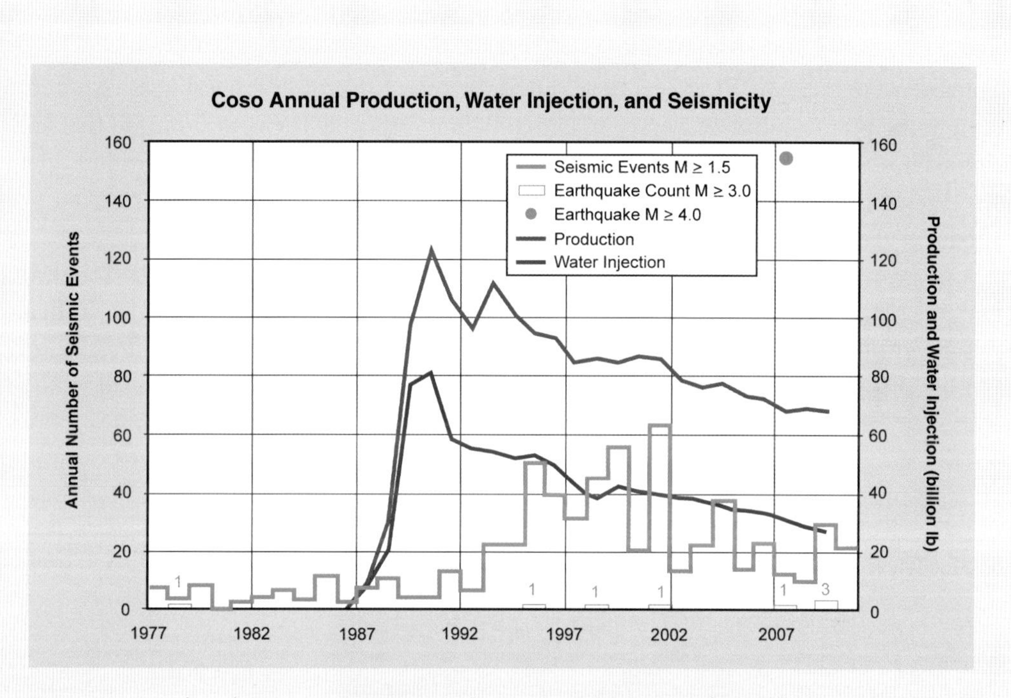

Figure 2 Annual production, water injection, and seismicity at the Coso geothermal field. SOURCE: Generated by the study committee from available data.

Enhanced Geothermal Systems

In addition to the vapor- and liquid-dominated resources already described, some regions have sufficiently high temperature at reasonably shallow depths for potential commercial development of EGS. To develop EGS some form of engineering is required to generate the permeability necessary in generally impermeable rocks to promote the circulation of hot water or steam for delivery to the surface at adequate rates to sustain operations. Previously referred to as "hot dry rock" projects, these systems are now referred to as "enhanced geothermal systems" or EGS (Figure 3.7).

The primary method employed to enhance rock permeability is hydraulic fracturing. This process, often termed "stimulation," requires the injection of a liquid at sufficient pressure in one well to overcome the confining pressures at depth and to thereby force open incipient fractures and planes of weakness or to create new fractures to allow fluids

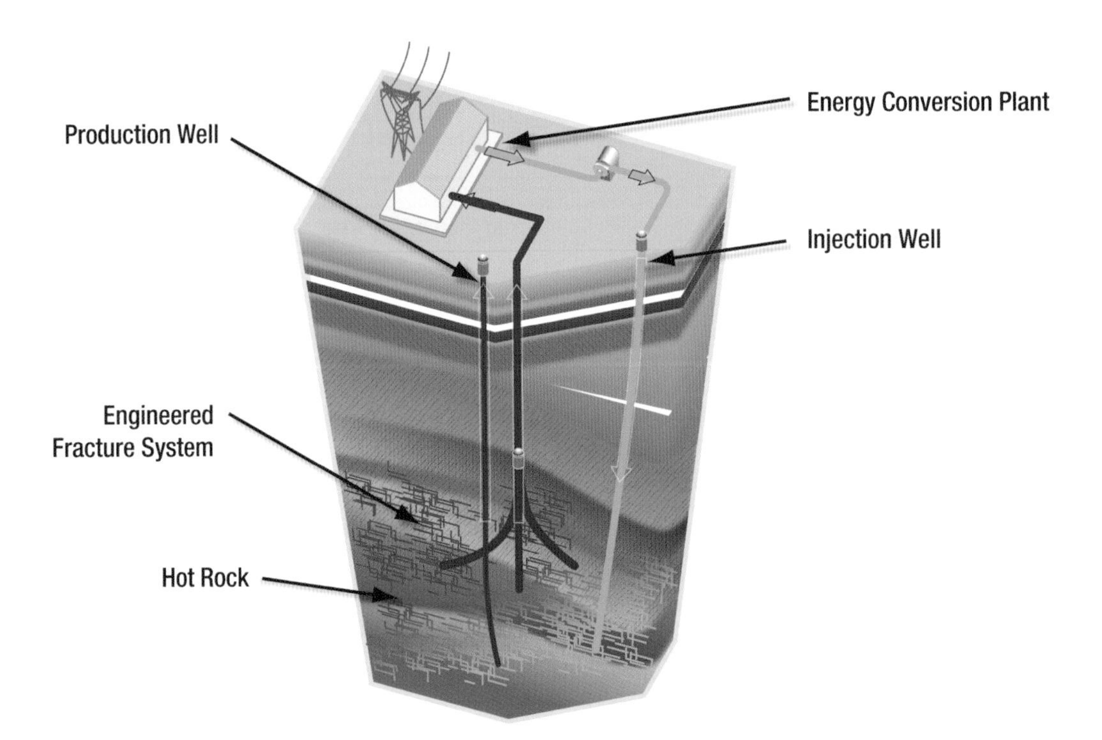

FIGURE 3.7 Schematic of an EGS development with an injection-production well pair and a power plant. The injection well (blue) is accompanied by a second (production) well (red) that is drilled to intersect the fractures generated by the injection well at a depth and appropriate lateral distance from the injection well. The distance allows the injected water to be sufficiently heated by the hot surrounding rock as it is circulated to the production well and pumped to the surface. Once at the surface the hot water can be flashed to steam or used to heat a secondary fluid that can be used in a binary cycle process. SOURCE: U.S. Department of Energy.

to flow more freely through the subsurface rock. The location of the new fractures can be determined by monitoring the microseismic response at the surface or downhole.

The history of the development of EGS projects in the United States began near the Los Alamos National Laboratory in New Mexico during the 1970s. That project provided a base for gaining experience in conducting hydraulic fracturing operations at high temperatures in low-permeability crystalline rocks. Data from this project have led to a series of similar EGS experiments in England, France, Germany, and Japan, followed more recently in Australia, Sweden, and Switzerland. In each case of active EGS development some induced seismicity has been registered. One recent example in Basel, Switzerland, generated an increased level of public awareness of the existence of induced seismicity (Box 3.3).

This Basel incident has become one of the best-known international induced seismic case studies, not because of local damage (which was minimal) but because of the immediate negative impact to the project due to the risk liability of induced seismicity. The urban setting for the project combined with the fact that this region is tectonically unstable and with a history of natural seismicity proved decisive in the project being terminated.

The occurrence of some post-shut-in seismicity at Basel and at another EGS project in Soultz-sous-Forêts, France, is a phenomenon that is not yet completely understood and can create added concern from the public standpoint in that some events are beyond the control of the operator. Understanding these post-shut-in events involves development of subsurface models with numerical simulations that can track the progress of the injected fluids through the rock and can calculate potential for further seismic activity. Development of coupled reservoir fluid flow and geomechanical simulation codes has been suggested as a way to advance this understanding (Majer et al., 2007) and may also have an impact on understanding post-shut-in phenomena related to other energy technologies (see also below).

CONVENTIONAL OIL AND GAS PRODUCTION INCLUDING ENHANCED OIL RECOVERY

In a conventional oil or gas reservoir, the reservoir rocks are generally pressurized above hydrostatic pressure due to compaction of sedimentary rocks over geologic time. The use of the term "reservoir" is common but may be misleading: the gas or oil does not exist in a single, large pool in the rocks, but in the pores of a rock formation. Compaction reduces the naturally occurring pore space in the rock (reduces the porosity) and either displaces reservoir fluids (hydrocarbons and water) or increases the pressure in the reservoir, or both. When penetrated by a well bore with the aid of pumping, fluids in the pressurized layer flow to the surface until the pressure in the reservoir is reduced to hydrostatic pressure. The reduction in pressure also causes gas to come out of the fluid, much like a bottle of soda when the cap is removed. The released gas can also help to drive the oil to the surface until the pressure is reduced to hydrostatic conditions.

BOX 3.3
Induced Seismic Activity in Basel, Switzerland

Basel, Switzerland, is in the southeastern region of the Upper Rhine Graben, a fault-bounded trough, and was selected as the site of a planned geothermal cogeneration plant. Basel is known to be an area of potential seismic risk but had not suffered a damaging earthquake since a **M** 6.2 earthquake in 1356 that destroyed much of the city. Due to awareness of historical seismicity, the geothermal project operators and planners had installed both borehole and surface seismic sensors that formed a network for monitoring any seismicity, whether natural or induced. The monitoring efforts included the drilling of six monitoring wells, ranging in depth from 300 m (~980 feet) to 2,750 m (~9,000 feet) in addition to a surface array of both weak and strong motion detectors. Recording of seismic activity began in early 2006 to record background seismicity.

The seismic monitoring arrays served several purposes. They recorded the background seismicity before well stimulation began and they were used to monitor the fracturing of the geothermal reservoir (the objective of the stimulation). Finally they could provide information (magnitude and location if possible) of any induced seismicity that might occur as a result of the stimulation. All monitoring stations were connected so that real-time data could be recorded and quickly analyzed.

The drilling of a deep geothermal well near the center of Basel (Figure 1) began in May 2006 and was completed some months later. Stimulation of the well to induce fractures for heat exchange with the geothermal source at 5,000 m (~16,400 feet) began on December 2 and was accompanied by a significant increase in the number of small seismic events (Figure 2). In accordance with the traffic light procedure—a procedure where increases in seismic activity beyond a certain, predetermined level trigger reactions by the operator to

Figure 1 Drilling activity in the middle of the city of Basel. SOURCE: KEYSTONE/Georgios Kefalas.

mitigate the occurrence of further events—injection was stopped in the early morning hours of December 8 after approximately 11,500 m^3 (~3 million gallons) of water were injected (Deichmann and Giardini, 2009) and after the recording of **M** 2.6 and **M** 2.7 seismic events. During this injection period, more than 10,500 seismic events were recorded (Häring et al., 2008). While the well was shut in (operations terminated), seismic activity continued, so it was decided to "bleed off" the pressure (reduce pressure through controlled release). On December 8, an earthquake of **M** 3.4 occurred in Basel and was clearly felt by the local population. This was followed by three more events greater than **M** 3.0. The project, operated by Geopower Basel AG as a partnership of both public and private companies, was immediately suspended and then ultimately abandoned almost 3 years later following further study and risk evaluation after these seismic events. However, increased seismicity activity over historical levels is likely to continue for 7 to 20 years based on Bachmann et al.'s (2009) model for induced seismicity.

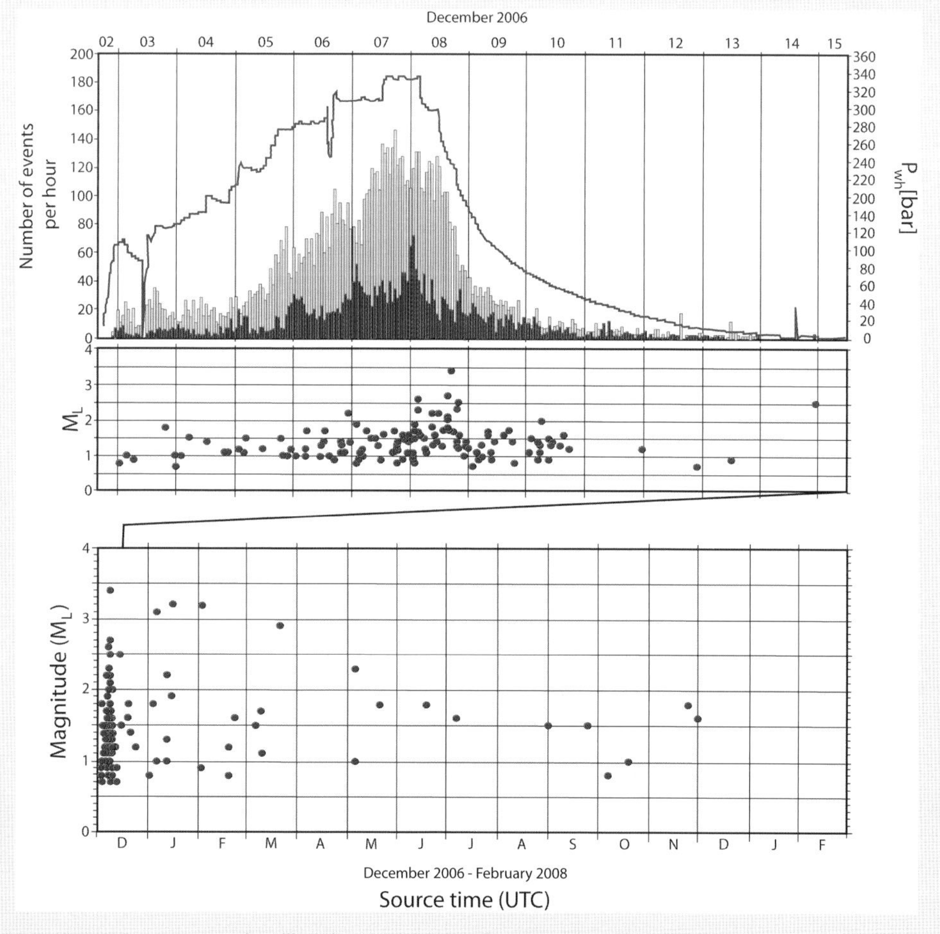

Figure 2 Seismic events and wellhead pressure at Basel. SOURCE: Kraft et al. (2009).

Flowing and pumped wells are considered "primary recovery" from the well and about 12 to 20 percent of the original oil in place in the reservoir is recovered in this manner. This relatively low rate of recovery results from several factors: (1) the decrease in natural reservoir pore pressure over time; (2) the natural porosity and permeability of the rock formation (which is an indication of how easily the oil can move through the formation to the well bore); and (3) the viscosity of the oil, which, when combined with porosity and permeability, is also an indicator of the ease with which oil can migrate through the rock. Recovery rates for natural gas are generally higher than for oil (up to 50 to 80 percent may be recovered through primary production methods) because gas expands naturally upon release of pressure and has a lower viscosity than liquid petroleum, contributing to the natural movement of gas up the well bore (Shepherd, 2009).

When primary recovery is no longer viable, petroleum companies may use a variety of technologies to extract the remaining oil and gas. These technologies include what are termed secondary and tertiary recovery methods; tertiary recovery is generally also referred to as enhanced oil recovery (EOR) (Shepherd, 2009). Figure 3.8 shows the differences between primary, secondary, and tertiary recovery methods.

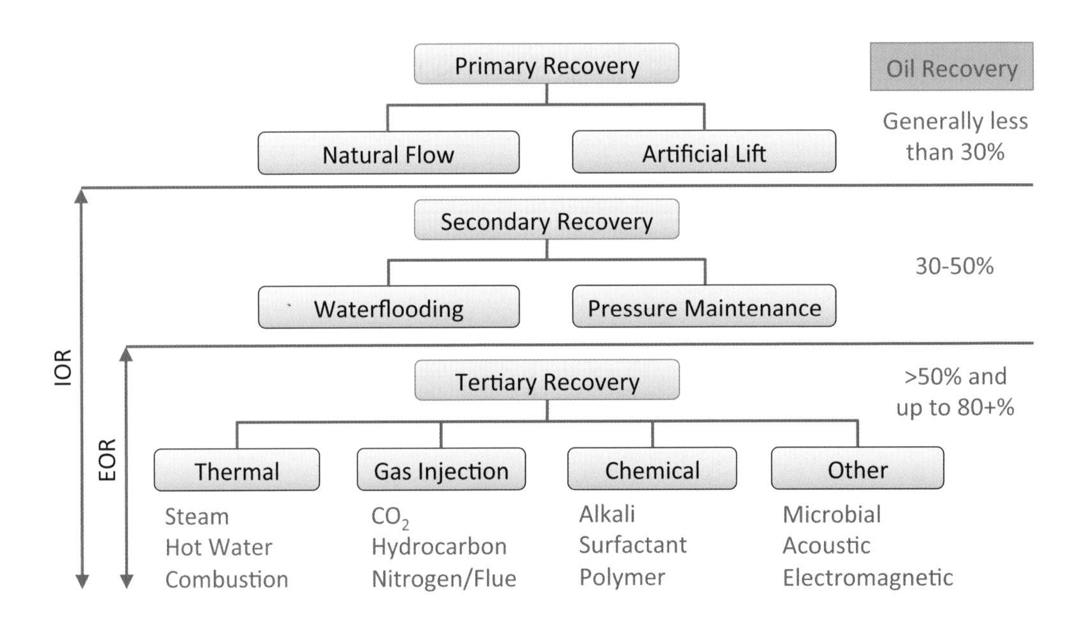

FIGURE 3.8 Schematic showing the progression of oil production from primary to tertiary recovery. IOR, improved oil recovery; EOR, enhanced oil recovery. SOURCE: Al-Mutairi and Kokal (2011).

Primary Oil and Gas Production

Although felt seismic activity known to be related to primary petroleum production is uncommon relative to the large number of operating oil and gas fields worldwide, withdrawal (extraction) of oil and gas has been linked to felt seismic events at 38 sites globally, 20 of which were in the United States (Appendix C; Box 1.1). These have included events in Texas, Oklahoma, California, Louisiana, Illinois, and Nebraska, the majority of which have been of **M** < 4.0 (Appendix C; see also Chapter 1); the well-documented events at the Lacq gas field in southwestern France (see Box 2.5); and the large events in the Gazli gas field in Uzbekistan (Box 3.4). Withdrawal of oil or gas from the subsurface can result

BOX 3.4
Induced Seismicity Related to Natural Gas Extraction: A Case from Gazli, Uzbekistan

The Gazli gas field is located about 500 miles (800 km) east of the Caspian Sea in a generally aseismic region of Uzbekistan. The gas deposits were discovered in 1956 and gas production began in 1962. The gas field lies within a large (38 km by 12 km [22.8 mile by 7.2 mile]) asymmetrical anticline over crystalline rocks. Large volumes of water were injected between 1962 and 1976 to enhance production, but subsidence and reduced gas pressures were reported despite this injection; the initial pressure in the gas field of about 70 atm (~71 bars or 1030 psi) in the 1960s decreased to about 30-35 atm (~30.4-35.5 bars or 435-515 psi) by 1976 and to about 15 atm (15.2 bars or 218 psi) by 1985. This pressure decrease indicates a net removal of mass, even with injection of large volumes of water. Thus, although the field operators had begun to use secondary recovery techniques (waterflooding), the cause of the earthquakes is attributed to pressure decrease due to fluid withdrawal.

On April 8, 1976, a **M** ~ 7 earthquake occurred about 20 km (12 miles) north of the gas field boundary. This was followed by another **M** ~ 7 earthquake on May 17, 1976. A third large earthquake (also **M** ~ 7) occurred on March 20, 1984. All three earthquakes had epicenters 10-20 km (6-12 miles) north of the gas field boundary, over an east-west distance of about 50 km (30 miles). Reported hypocentral depths of these large earthquakes were 10-15 km (6-9 miles). Geodesic measurements indicated surface uplift of some 70-80 cm (~28 to 31.5 inches) north of the gas field at the epicentral locations of the three large earthquakes; this uplift is consistent with thrust movement on faults dipping to the north. However, source modeling indicates that the ruptures progressed downward, which is uncommon for thrust mechanism earthquakes. The locations and magnitudes of these large earthquakes were determined from worldwide seismographic data and are therefore somewhat uncertain, leading to some uncertainty on the causal relationship between gas extraction and earthquake activity. Nonetheless, observations of crustal uplift and the proximity of these large earthquakes to the Gazli gas field in a previously seismically quiet region strongly suggest that they were induced by hydrocarbon extraction.

SOURCES: Adushki et al. (2000); Grasso (1992); Simpson and Leith (1985).

in a net decrease in pore pressure in the reservoir over time, particularly if fluids are not reinjected to maintain or regain original pore pressure conditions (see also other technology descriptions, below, and Chapter 2). This change in pore pressure can cause changes in the state of stress of the surrounding rock mass and of nearby faults, with the potential to result in induced seismic events.

Secondary Oil and Gas Recovery

Secondary recovery is the process of injecting water (often described as a "waterflood") or gas (also known as pressure maintenance) into a petroleum reservoir. The water or gas replaces the produced hydrocarbons and water in order to maintain the reservoir pressures and is used to "sweep" an oil reservoir; injected gas may become dissolved in the oil, reducing the oil's viscosity. Secondary recovery processes drive hydrocarbons trapped in the rocks from the injection well toward production wells (Shepherd, 2009; Figure 3.9). Waterflood or pressure maintenance projects can result in recovery of up to 40 percent of the initial petroleum in the reservoir (DOE, 2011). The number of permitted wells that use

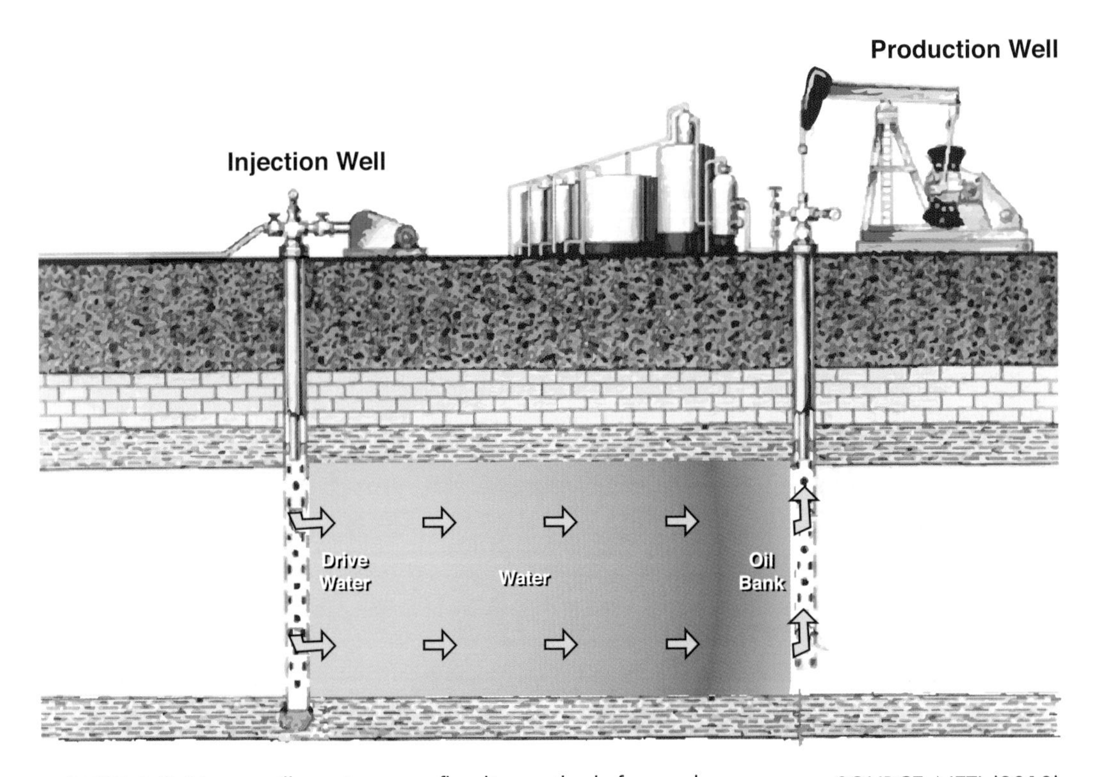

FIGURE 3.9 Diagram illustrating waterflooding method of secondary recovery. SOURCE: NETL (2010).

waterflooding in the United States is about 108,000; in Texas alone, current data from the Railroad Commission of Texas indicate that more than 36,000 wells are currently permitted to use saltwater injection for the purposes of secondary recovery.[4]

Injection pressures and volumes in waterflooding projects are generally controlled to avoid increasing the pore pressure in the reservoir above the initial reservoir pore pressure. Nonetheless, reservoir pore pressure can increase as a result of waterflooding, and felt induced seismic events at 27 sites globally (18 of which have been in the United States) have been caused by or likely related to waterflooding (Chapter 1, Box 1.1; Appendix C). Waterflooding at the Rangely Field in Colorado induced seismic events with magnitudes up to **M** 3.4 (Chapter 2, Box 2.4). Near Snyder, Texas, seismic events with magnitudes as large as **M** 4.6 occurred in 1978 after the initiation of a large (25 million barrel per year [10.2 trillion gallons per year]) waterflooding project in Cogdell Field (Davis and Pennington, 1989; Nicholson and Wesson, 1990; see also Appendix C).

Tertiary Oil and Gas Recovery (EOR)

Tertiary recovery is the process of recovering greater amounts (often greater than 50 percent) of the original oil and gas contained in a reservoir (DOE, 2011) and is generally, though not exclusively, initiated after the use of secondary recovery operations.[5] In addition to maintaining reservoir pore pressure, EOR methods help displace the hydrocarbons toward the production well. These methods can be broadly grouped into three main categories: thermal, miscible displacement, and chemical injection (polymer flooding) (Shepherd, 2009). Chemical injection methods are primarily used in California but are not commonly used elsewhere in the United States and are not discussed further. Note also that "other" methods in Figure 3.8 include microbial, acoustic, and electromagnetic methods; these are not frequently used and are not discussed further.

Thermal techniques change the viscosity of oil in the reservoir by heating it through the injection of steam or air (Shepherd, 2009). Heating lowers the viscosity of the fluid and allows hydrocarbons to flow more easily through a reservoir toward a production well. Over 40 percent of EOR operations in the United States use this method; it is most commonly employed in fields with high-viscosity oils (DOE, 2011). Miscible displacement is generally used for lower-viscosity oils and involves injecting gases such as nitrogen or CO_2 that can reduce the viscosity of the oil and physically displace it toward production wells (Figure 3.10). Nearly 60 percent of EOR projects in the United States use this gas injection technique (DOE, 2011). In the United States, over 600 million tons of CO_2 (11 trillion standard cubic feet; ~540 million metric tonnes) have been injected in ~13,000 wells for

[4] See www.rrc.state.tx.us/data/wells/fluids.php.

[5] See www.glossary.oilfield.slb.com/Display.cfm?Term=enhanced%20oil%20recovery.

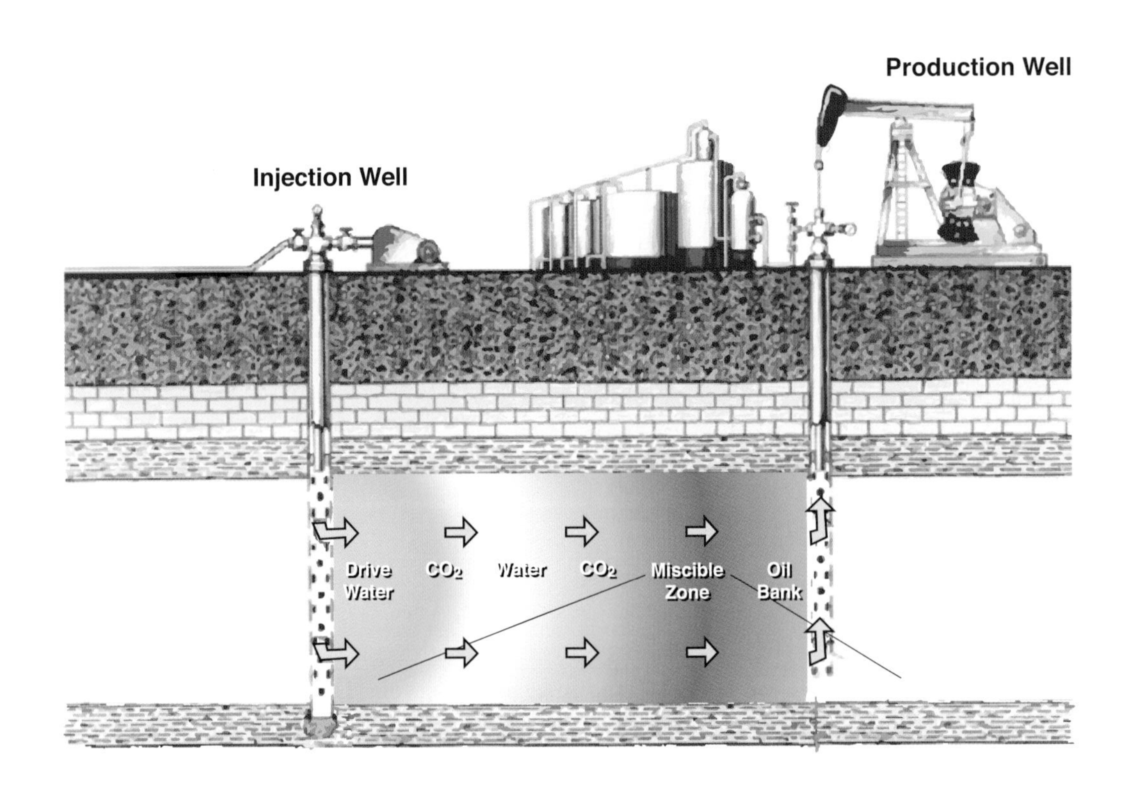

FIGURE 3.10 Enhanced oil recovery through CO_2 injection. SOURCE: NETL (2010).

EOR as of 2007 (Meyer, 2007). Current records from the Railroad Commission of Texas indicate that more than 9,400 wells are permitted in Texas alone for CO_2 injection for EOR.[6] Among the many thousands of wells used for EOR in the United States, the committee did not find any documented instances of felt induced seismicity in the published literature or from experts in the field with whom the committee communicated during the study.

[6] See www.rrc.state.tx.us/data/wells/fluids.php.

One reason for the apparent lack of induced seismicity with EOR may be that EOR operations routinely attempt to maintain the pore pressure within a field at levels near preproduction pore pressures. This "balance" of the pore pressure means only a minimum pressure change occurs in the reservoir, reducing the possibility of induced seismic events; this maintenance of pore pressure is achieved broadly by maintaining balance between the amount of fluid being injected and the amount being withdrawn. EOR using CO_2 injection is also considered one form of CCS, a technology under broader development in several other geological settings as part of the effort to reduce greenhouse gas emissions. CCS is discussed in detail later in this chapter.

UNCONVENTIONAL OIL AND GAS PRODUCTION INCLUDING SHALE RESERVOIRS

The permeability of rock in the subsurface varies tremendously (see Figure 2.1). Mudstone, siltstone, or shale formations that are high in organic content may contain significant amounts of natural gas and oil but have very low permeability; a shale formation that contains predominantly gas and/or oil is called a shale reservoir. Shales that are actively drilled for both oil and gas development in the United States are, for example, the Barnett, Marcellus, Eagle Ford, and Bakken formations (Figure 3.11).

Unlike conventional oil and gas fields, where the hydrocarbons were formed in source rocks high in organic content and then migrated over geologic time into porous rock such as sandstones and limestones that serve as the reservoirs today, the hydrocarbons in shales have developed from and remained for the most part trapped in their original source rock (organic-rich fine-grained sediments) because of the very low permeability of the shales. The shale gas resides in the microporosity in the shale layers and is held in place by a combination of cap rock, adsorption of gas onto the shale grains, and low permeability. The last of these effects is primarily responsible for the low production rates of drilled shales before being hydraulically fractured. Hydraulic fracturing creates additional pathways among the micropores for the gas to flow to the wellbore (see, e.g., NRC, 1996). This type of hydrocarbon reservoir, which requires additional engineered solutions for extraction of hydrocarbons, is often called an unconventional reservoir.

Extraction of gas and oil from these unconventional reservoirs has been made feasible through the combined application of horizontal drilling and hydraulic fracturing, technologies developed by the petroleum industry and through research supported by the Department of Energy (EIA, 1993, 2011; NETL, 2007; NRC, 2001). Hydraulic fracturing has been used for over 50 years to stimulate some conventional reservoirs (EIA, 2011) but is required to produce from low-permeability reservoirs such as shales for which commercially viable technology was developed by Mitchell Energy during the 1980s and 1990s (EIA, 2011). A large upswing in the use of horizontal drilling and hydraulic fracturing

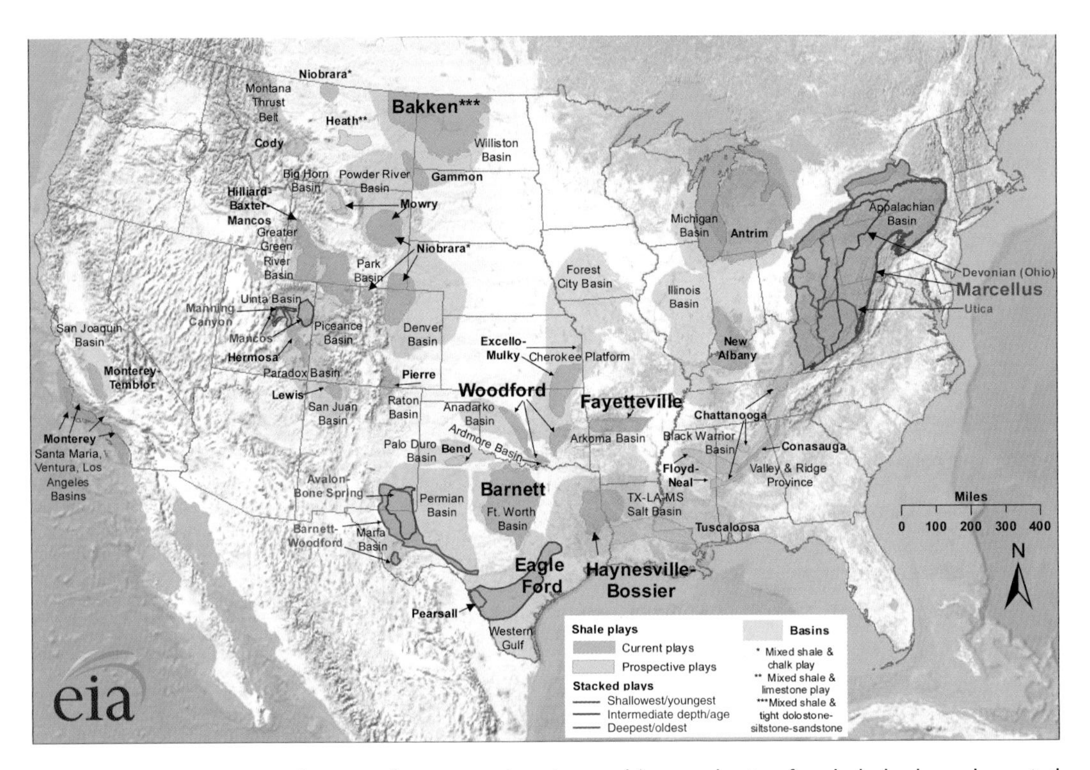

FIGURE 3.11 Location of areas of active exploration and/or production for shale hydrocarbons (oil and gas) in the contiguous United States. Light pink areas are major sedimentary basins; dark pink areas (e.g., Eagle Ford, Barnett) are under active development and production for gas or oil from shale; orange areas are prospective regions currently being explored for potential oil or gas development from shale. Several shale units of different ages may overlie one another, and these units are outlined in thick red, blue, and purple lines representing youngest to oldest shale units, respectively. A "play" is a set of oil or gas accumulations that share similar geologic, geographic, and time characteristics. SOURCE: EIA (2011). Available at www.eia.gov/pub/oil_gas/natural_gas/analysis_publications/maps/maps.htm.

occurred in the late 1990s and continues to the present day; estimates suggest that today approximately 60 percent of the wells drilled are hydraulically fractured (Montgomery and Smith, 2010).

A typical production well in shale is drilled vertically to an appropriate depth and then turned horizontally to extend the well bore through the target shale formation. The horizontal segment (or "lateral") of the well typically extends over 1-2 miles (~1.8-3 km) (Box 3.5). To facilitate the flow of the gas or oil into the well bore, the permeability through the shale reservoir is increased by the creation of artificial fracture networks in the shale around the horizontal portion of the well bore through the process of hydraulic fracturing (Box 3.5). Microseisms generally of **M** < 0 are induced during a hydraulic fracture treatment, and the locations of these microseisms are used to help understand the location of the artificially created fractures and can be used as stress measurement tools (Appendix I describes this kind of microseismic monitoring; see also Engelder, 1993).

After the hydraulic fracturing is completed, a process known as flowback occurs. The well is opened and injected hydraulic fracture water is allowed to flow back from the formation into the well. For tight shale formations, between 10 and 50 percent of the hydraulic fracture water is returned (King, 2010). The flowback water may be reused as fracturing water for another hydraulic fracture procedure, may be disposed of in a wastewater injection well (see next section), may be stored, or may be treated to a purity that would allow for its safe release to the environment or for its use for other beneficial purposes. Two National Research Council reports (NRC, 2010, 2012) describe in some detail the potential options for management and beneficial use of wastewater from industrial activities.

The process of hydraulic fracturing a well as presently implemented for shale gas recovery does not pose a high risk for inducing felt seismic events (**M** > 2). Estimates suggest that over 35,000 wells for shale gas development exist in the United States today (EPA, 2011). Only one case has been documented worldwide in which hydraulic fracturing for shale gas development has been confirmed as the cause of felt seismic events. This event occurred in Blackpool, England, in 2011 (De Pater and Baisch, 2011; Box 3.6). Three other possible earthquake sequences have been discussed in the literature that may be associated with hydraulic fracturing in Oklahoma, only one of which was related to shale gas production. In the most recent case, in 2011, hydraulic fracturing for shale gas production was cited as the possible cause of felt induced seismic events, the largest of which was **M** 2.8 (Holland, 2011; Appendix J). The close proximity and timing of the earthquakes to the hydraulic fracturing well suggested a possible, but not fully established, link. However, the quality of the event locations was not adequate to fully establish a direct causal link to the hydraulic fracture treatment.

The two other possible cases in Oklahoma discussed by Nicholson and Wesson (1990) are listed under "Less Well Documented or Possible Cases" in their original paper (see also Appendix C). Both cases were associated in time with hydraulic fracturing related to stimulation of a conventional oil and gas field, not for shale gas production. The older of the two cases relates to a series of earthquakes that occurred on June 23, 1978, near the commercial stimulation of a 3,050-m (10,000-foot) well near Wilson, Oklahoma. Seventy earthquakes occurred in 6.2 hours (Luza and Lawson, 1980; Nicholson and Wesson, 1990). In the third case, two earthquakes were felt in a sequence in Oklahoma in May 1979, during the time that a well was vertically stimulated in three different zones, ranging from deep to shallow (ranging from 3,700 to 3,000 m depth [~12,000 to 10,000 feet]). The largest event in this third case was **M** 1.9. The well was located 1 km (3,280 feet) from a seismic monitoring station. The first hydraulic fracture treatment at 3,700 m depth was followed 20 hours later by about 50 earthquakes that occurred over a 4-hour time period. Forty earthquakes immediately followed the second hydraulic fracture treatment at 3,400 m, over a time period of 2 hours. No earthquakes were recorded during the third hydraulic fracture

BOX 3.5
Hydraulic Fracturing

A hydraulic fracture is a controlled, high-pressure injection of fluid and proppant into a well to fracture the target formation (see Figure). "Proppant" refers to sand or manmade ceramics used to keep the fractures open after fluid injection stops. The injected fluid is usually a combination of water and small amounts of chemical additives that reduce pipe flow friction, minimize rock formation damage, and help carry proppant into the fractures (see also Box 2.3; DOE, 2009; King, 2012). Horizontal wells are hydraulically fractured in multiple pumping "stages," starting at the far end of the horizontal well and progressing toward the wellhead. Each fracture stage is isolated within the horizontal well with packers or mechanical sleeves that open and close each zone. After the entire hydraulic fracture procedure is completed, the injected fluid is allowed to flow back into the well, leaving the proppant in the newly created fractures. The amount of fracturing fluid used in one horizontal well fracturing stage varies, depending in large part on the geologic formation, and is on the order of millions of gallons per well. Generally, water volumes are estimated from 2 to 5.6 million gallons per well (DOE, 2009; King, 2012; Nicot and Scanlon, 2012; Soeder and Kappel, 2009). Horizontal wells can be hydraulically fractured in one to more than 30 stages depending on the length of the horizontal well.

The distance and direction of the manmade fractures propagating from the well vary depending on the type of hydraulic fracture treatment and the geologic properties near the well, including the rock toughness and stress state in the formation. In general, the fractures are observed from geophysical surveys such as microseismic (Appendix I) and tiltmeters (Cipolla and Wright, 2002) to propagate perpendicular to the direction of the minimum in situ stress. The induced fractures can form a complex fracture network in areas of low horizontal stress differences or simple fracture geometry in higher differential stress areas. Although the extent and direction of the fractures are not known precisely, hydraulic fractures may extend on the order of one hundred to over a thousand feet from the well. The upward growth of the hydraulic fracture tends to be limited by the horizontal layering (bedding) of the shale formations and by the vertical stress exerted by overlying rock and rarely extends up more than a few hundred feet (less than 100 m) from the wellbore (Fisher, 2010; Fisher and Warpinski, 2011). The geometry of hydraulic fractures can be estimated using a special seismic monitoring technique termed microseismic mapping (see Appendix I), although this geophysical procedure is completed on only a small percentage of hydraulically fractured wells, largely due to the cost.

BOX 3.6
Felt Earthquakes Near Blackpool, England, Related to Hydraulic Fracturing

Hydraulic fracturing of the Preese Hall-1 well in the Blackpool area of England caused seismicity in April (**M** 2.3) and May (**M** 1.5) 2011. The April earthquake was felt in northern England and was widely reported in the press. The well was drilled and hydraulically fractured by Cuadrilla Resources to explore the gas potential of the Bowland Shale Formation.

The Preese Hall-1 exploration well was stimulated vertically to 9,004 feet measured depth with five hydraulic fracture stages. The April **M** 2.3 event occurred during stage 2, and the May **M** 1.5 occurred during stage 4; in

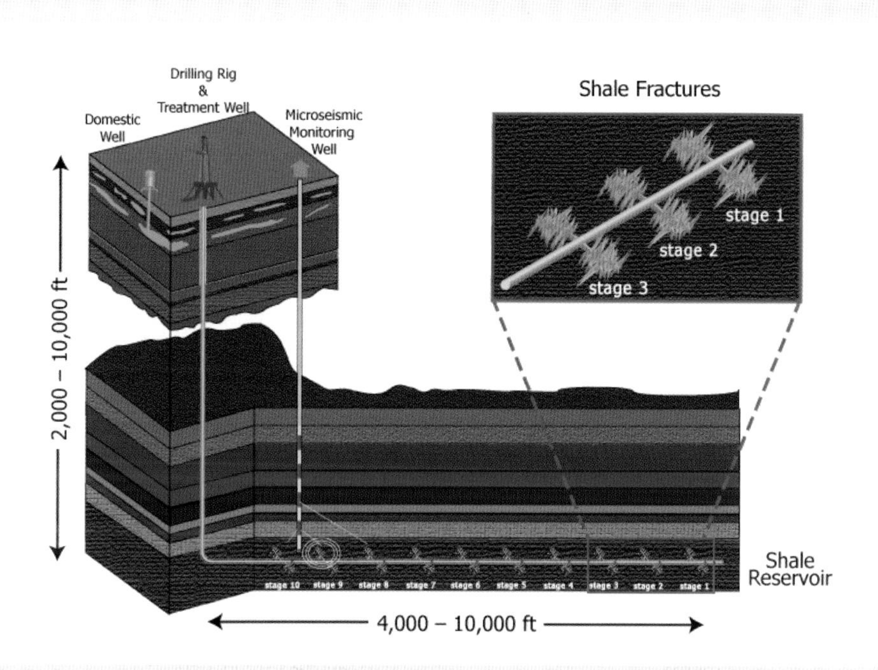

Figure Schematic diagram of a horizontal well following a 10-stage hydraulic fracture treatment. Upper right inset shows a magnified view of the induced fractures (yellow) created during the hydraulic fracture treatment. The relative depths of local water wells is shown near the surface for scale, labeled "domestic well." The formation depth and horizontal well length vary from area to area; the depth and well length numbers shown are approximate averages for North America. The well is fractured in stages from the end of the well (stage 1) to the start of the well (stage 10). Each hydraulic fracture stage is isolated within the wellbore as discussed in the text. Depths and distances of 2,000-10,000 feet correspond to about 600-3,000 m. SOURCE: Adapted after Southwestern Energy, used with permission.

addition approximately 50 weaker events were detected after additional seismic stations were deployed (De Pater and Baisch, 2011). Cuadrilla Resources initiated an extensive study of the incident, including installing portable seismic stations and a detailed seismic analysis as well as geomechanical studies and core studies, which were released to the public on their website. The research demonstrates that the hydraulic fracturing induced the seismic events. A report by Geosphere Ltd. (Harper, 2011) suggests the propagation of the fracturing fluid and pressure went farther than expected along the bedding planes. A nearby, apparently unstable fault was reactivated by the increase in fluid pressure, which caused the seismic events (Harper, 2011).

SOURCES: De Pater and Baisch (2011).

treatment at 3,000 m. All three Oklahoma cases demonstrate a reoccurring problem in induced seismicity studies: the seismic events are small, the regional networks are sparse, and the data quality is often too poor to fully confirm a causal link to fluid injection for energy development (see also Chapter 1).

INJECTION WELLS USED FOR THE DISPOSAL OF WATER ASSOCIATED WITH ENERGY EXTRACTION

In addition to fluid injection for specific kinds of energy development (e.g., water injection to produce steam for geothermal energy recovery, or fluid injection for waterflooding [secondary recovery]), water injection to dispose of water generated as a result of geothermal and oil and gas production operations is very common in the United States. Water that must be disposed of originates from production (see, e.g., NRC, 2010) or from flowback. Hereafter we refer to this kind of water broadly as wastewater; Chapter 4 clarifies the different kinds of water from energy production that are disposed of and the different classes of wells that are designated in the United States for this purpose. A recent study by Argonne National Laboratory estimated the total oil and gas fluid recovered from flowback after hydraulic fracturing operations and waste fluid produced during daily oil and gas production in the United States to be 20.9 billion barrels (about 878 billion gallons) of water per year (Clark and Veil, 2009). The majority (95 percent) of this water was managed through underground injection and more than half (55 percent) was injected for the purpose of enhanced recovery (Clark and Veil, 2009) (see the section Tertiary Oil and Gas Recovery [EOR] in this chapter). Just over one-third of the total wastewater volume (39 percent) or 6 billion barrels (252 billion gallons) was injected in disposal wells. Table 3.2 shows the water volumes produced in conjunction with oil and gas operations for various states. Importantly, other types of fluid may also be disposed of through underground injection (industrial wastes, for example, from manufacturing unrelated to energy production); these different kinds of underground injection are also discussed in Chapter 4.

The annual volume of wastewater in the United States is disposed of in many tens of thousands of injection wells. For example, in Texas, over 50,000 Class II[7] injection wells were permitted as of 2010 (of which approximately 40 percent would be associated with disposal of wastewater and the remainder associated with waterflooding for secondary recovery; Texas RRC, 2010) (Figure 3.12).

Felt induced seismicity potentially related to Class II water injection wells has been identified at individual sites in Arkansas (see Chapter 4), Ohio, and Texas (Box 3.7). USGS

[7] Wells in the Environmental Protection Agency's (EPA's) Underground Injection Control (UIC) program are described and regulated under one of six "classes." Class II wells are specifically those that address injection of brines and other fluids associated with oil and gas production and hydrocarbons for storage. EPA's well class system and the UIC program are described in more detail in Chapter 4.

TABLE 3.2 U.S. Onshore and Offshore Oil, Gas, and Produced Water Generation for 2007

State	Crude Oil (bbl/year)	Total Gas (Mmcf)	Produced Water (bbl/year)	Data Source
Alabama	5,028,000	285,000	119,004,000	1
Alaska	263,595,000	3,498,000	801,336,000	1
Arizona	43,000	1,000	68,000	1, 2
Arkansas	6,103,000	272,000	166,011,000	2, 3
California	244,000,000	312,000	2,552,194,000	2, 3
Colorado	2,375,000	1,288,000	383,846,000	1,3
Florida	2,078,000	2,000	50,296,000	1
Illinois	3,202,000	No data	136,872,000	1, 5
Indiana	1,727,000	4,000	40,200,000	1, 2
Kansas	36,612,000	371,000	1,244,329,000	1, 2
Kentucky	3,572,000	95,000	24,607,000	1, 3, 6
Louisiana	52,495,000	1,382,000	1,149,643,000	1
Michigan	5,180,000	168,000	114,580,000	1, 3
Mississippi	20,027,000	97,000	330,730,000	1
Missouri	80,000	No data	1,613,000	1
Montana	34,749,000	95,000	182,266,000	1
Nebraska	2,335,000	1,000	49,312,000	1
Nevada	408,000	0	6,785,000	1, 2
New Mexico	59,138,000	1,526,000	665,685,000	1
New York	378,000	55,000	649,000	2
North Dakota	44,543,000	71,000	134,991,000	2, 4
Tennessee	5,422,000	86,000	6,940,000	1, 2
Texas	60,760,000	1,643,000	2,195,180,000	2, 6
Utah	1,537,000	172,000	3,912,000	3
Virginia	1,665,000	12,000	4,186,000	1, 2
West Virginia	350,000	1,000	2,263,000	4, 6
Wyoming	342,087,000	6,878,000	7,376,913,000	3, 4
State Total	**1,273,759,000**	**21,290,000**	**20,258,560,000**	
Federal Offshore	467,180,000	2,787,000	587,353,000	1
Tribal Lands	9,513,000	297,000	149,261,000	2, 6
Federal Total	**476,693,000**	**3,084,000**	**736,614,000**	
U.S. Total	**1,750,452,000**	**24,374,000**	**20,995,174,000**	

NOTE: 1, provided directly to Argonne by state agency; 2, obtained via published report or electronically; 3, obtained via electronic database; 4, obtained from website in form other than a published report or electronic database; 5, obtained from EIA; 6, produced water volumes are estimated from production volumes.
SOURCE: Clark and Veil (2009).

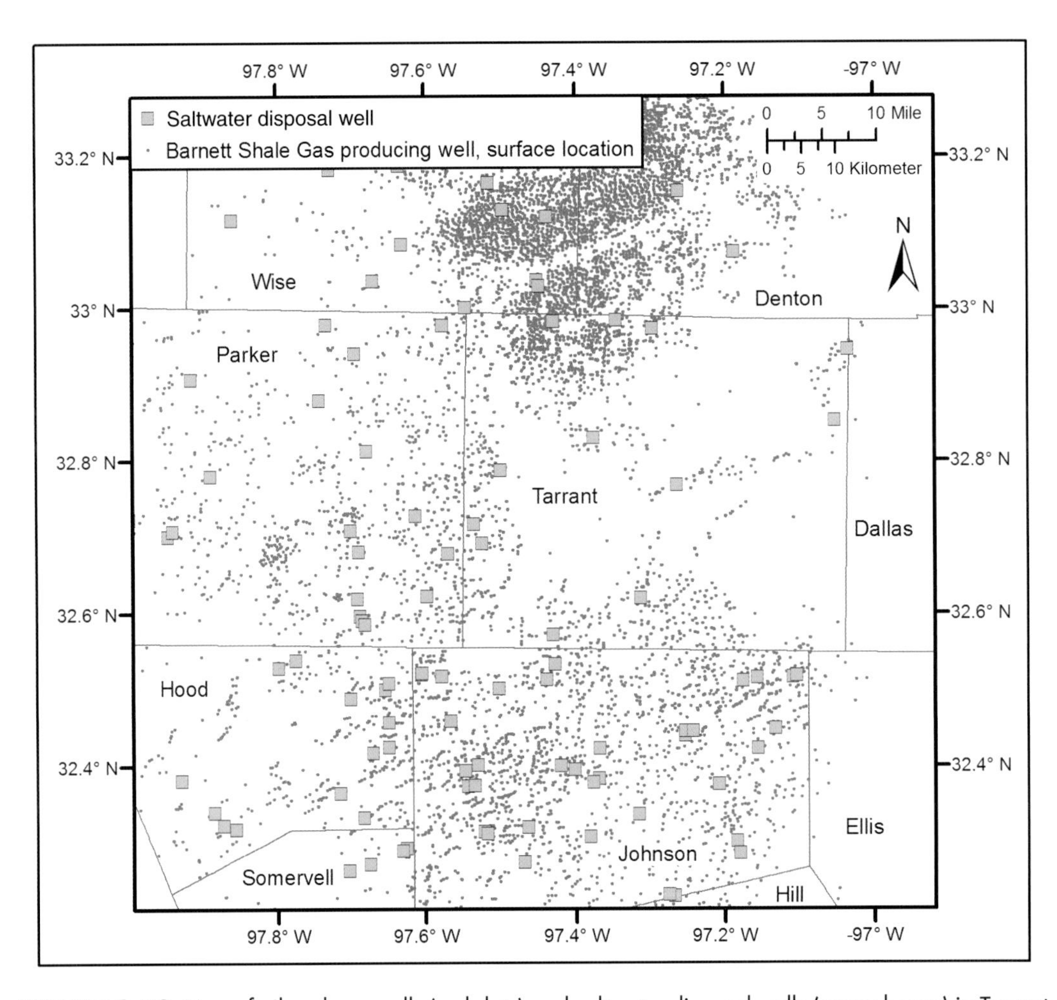

FIGURE 3.12 Map of oil and gas wells (red dots) and saltwater disposal wells (green boxes) in Tarrant and surrounding counties in Texas. The approximate location of the Dallas-Fort Worth (DFW) airport is marked with box (as labeled), along with the injection wells near the airport. SOURCE: Modified from Frohlich et al. (2010).

researchers are investigating whether a recent increase in the rate of **M** > 3.0 earthquakes in the state of Oklahoma (see Figure 3.13) might be attributed to wastewater injection (Ellsworth et al., 2012). One of the best-documented cases of induced seismicity from fluid injection is in the Paradox Basin, Colorado, where brine from a natural seep has been reinjected in one disposal well at 14,000 to 15,000 feet (4,300 to 4,600 m) depth since 1996 to prevent brine flow into the Colorado River (Appendix K). To date over 4,600 induced seismic events (**M** 0.5 to **M** 4.3) as far away as 16 km (9.9 miles) from the injection well have been documented in the Paradox Basin (Block, 2011). Although the number of felt induced seismic events relative

BOX 3.7
Dallas–Fort Worth Earthquake Swarm October 2008 to May 2009

A series of **M** 2.5 to M 3.3 earthquakes occurred in the Dallas–Fort Worth (DFW) area of Texas, where earthquakes were felt and reported by local residents in October 2008 and May 2009. The National Earthquake Information Center (NEIC) located the earthquakes in the vicinity of the DFW airport.

The state of Texas historically experienced a low rate of natural seismicity at the time of these earthquakes and the entire state has only two permanent seismographic stations operated by the NEIC. Because of the sparse seismographic station coverage, the NEIC can only locate events in Texas that are greater than about **M** 2.5 with location accuracy of plus or minus 6 miles or 10 km. Researchers from the University of Texas (UT) and Southern Methodist University (SMU) deployed a temporary network of six seismographic stations in the DFW area to locate seismic events more precisely. The UT-SMU seismic array ran from November 9, 2008, to January 2, 2009, and located 11 earthquakes that spanned a 1-km-long, north-south trending zone in close proximity to a saltwater disposal (SWD) well used for wastewater injection by Chesapeake Oil and Gas Company. The wastewater originated from wells in the vicinity of the DFW airport producing from the Barnett Shale (Figure 3.11). The first felt DFW earthquakes started about 6 weeks after injection into the disposal well was initiated. The close correspondence of the earthquakes with the location and depth of the well, together with the close timing of the start of injection and the start of seismic activity, strongly suggest that injection was the cause of the seismic activity.

A state tectonic map compiled by the Texas Bureau of Economic Geology shows a northeast trending normal fault in the subsurface in close proximity to the SWD injection well. The earthquake swarm continues in the DFW area to this date, with **M** 2.6 or less events occurring prior to August 2011, over 2 years after shutdown of the injection well (Eisner, 2011). The persistent seismicity after the nearby injection wells were shut in demonstrates the difficulty in assessing whether the seismic activity is induced or natural. Similar to the post-shut-in events that have occurred in relation to EGS projects in France and Switzerland, understanding the cause and magnitude of these events through time requires further research that combines field observations and data with fluid flow and geomechanical simulation codes.

Box continues

BOX 3.7 Continued

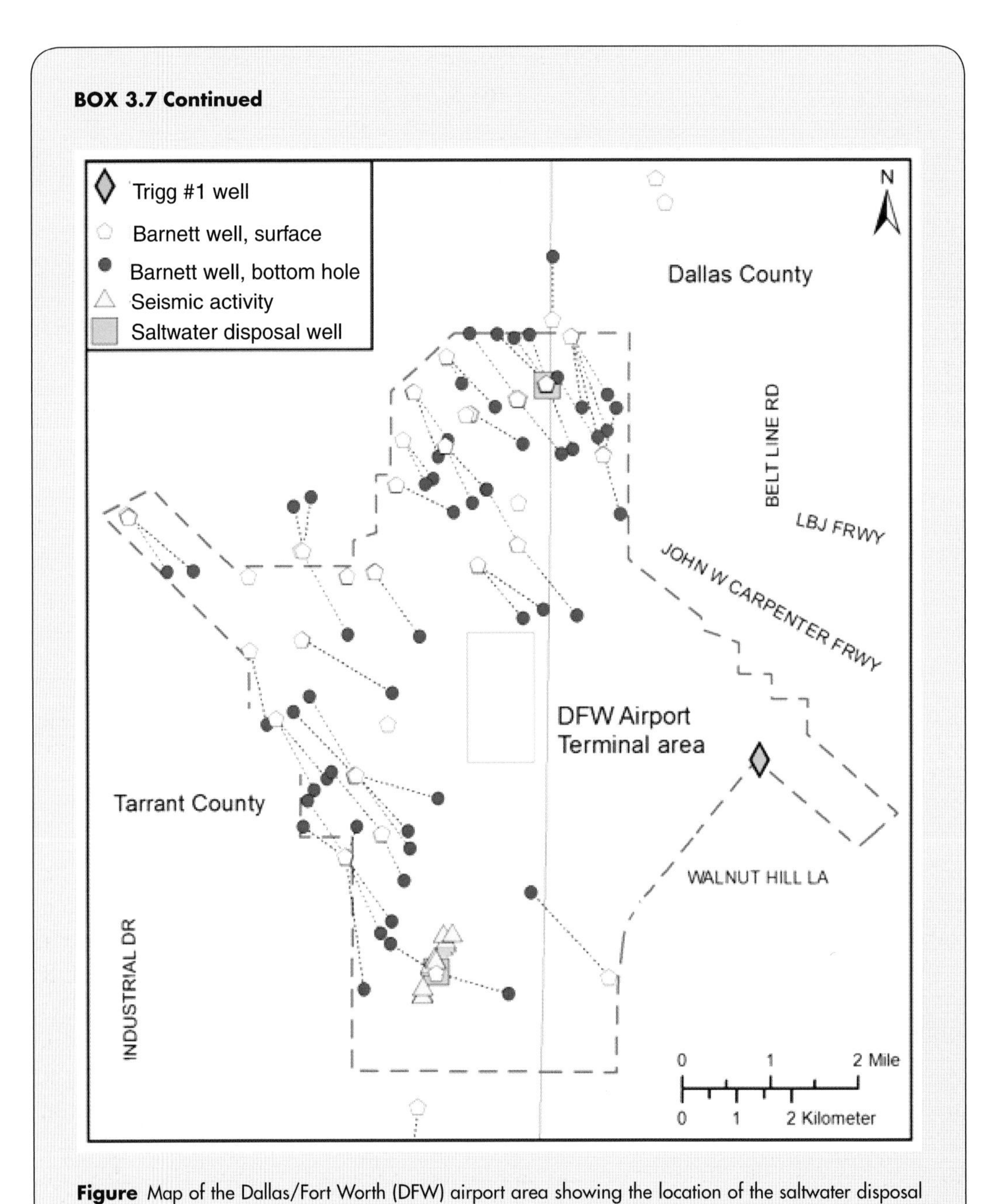

Figure Map of the Dallas/Fort Worth (DFW) airport area showing the location of the saltwater disposal well and the location of the earthquakes. SOURCE: Frohlich et al. (2010).

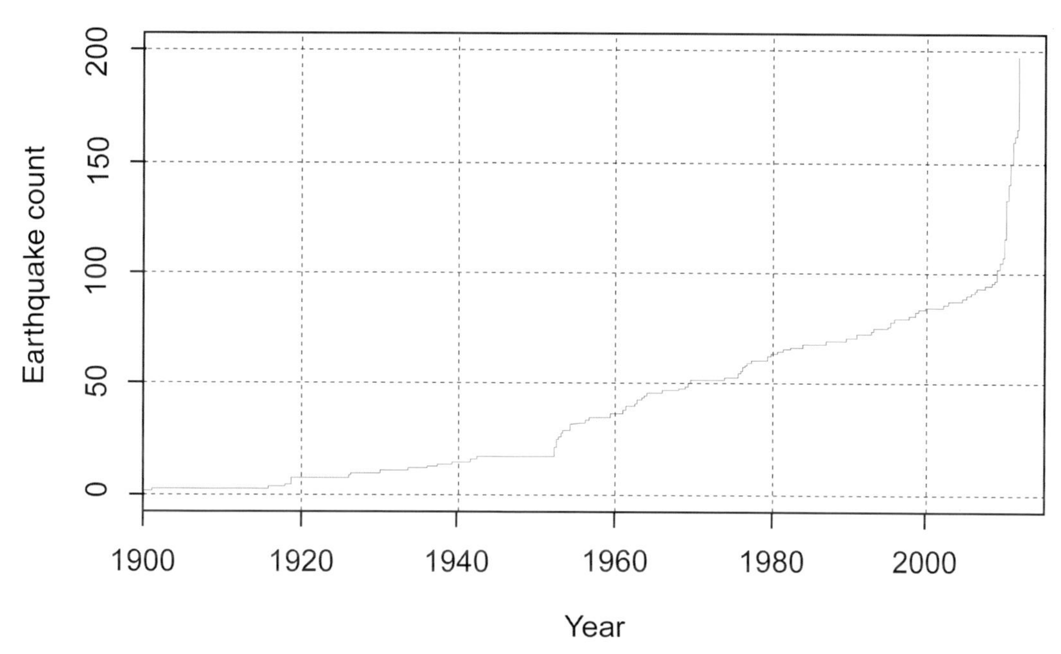

FIGURE 3.13 Graph showing the cumulative number of earthquakes **M** > 3.0 in the central Oklahoma region (34-37°N, 94-100°W) from 1900 to present day, showing a dramatic but as yet unexplained increase in seismicity since 2009. SOURCE: Ellsworth et al. (2012).

to the tens of thousands of produced water injection wells is small, the events themselves can cause considerable public concern. Addressing the causes and conditions for these events is useful for understanding induced seismicity potential for future wastewater injection projects.

Water injection wells only inject (dispose of) fluid, in contrast to injection wells for EOR or liquid-dominated geothermal systems where the fluid injected is approximately equivalent to the fluid extracted. Fluid injection in proximity to a favorably oriented fault system with near-critical stresses has an increased potential to generate felt induced seismic events in the absence of nearby extraction that could help maintain reservoir pressure. Class II injection wells used only for the purpose of water disposal normally do not have a detailed geologic review performed, and often data are not available to make such a review. Thus, although fluid pressure in the injection zone and the fracturing pressure of the injection zone can be measured after the disposal well is drilled, the location of possible faults is often not known as part of standard well siting and drilling procedures. Importantly, the mere presence of a fault does not always correlate to increased potential for induced seismicity. Chapter 6 discusses potential steps toward best practices with these challenges in mind.

CARBON CAPTURE AND STORAGE

Introduction of large amounts of CO_2, a greenhouse gas, into the atmosphere is considered a likely driver in climate change (NRC, 2011). In 2010 approximately 33.5 billion metric tonnes of CO_2 (~37 million tons) were introduced to the atmosphere by industry, transportation, and agricultural production globally (Boden and Blasing, 2011; Friedlingstein et al., 2010). For a number of years research has explored various methods for reducing carbon emissions to the atmosphere, including methods that can capture CO_2 from point sources (e.g., fossil fuel burning power plants, industrial plants, and refineries), transport it to a geological storage site, and inject it into the ground for permanent storage (sometimes called sequestration) and monitoring (shown schematically in Figure 3.14). If successful and economical, CCS could become an important technology for reducing CO_2 emissions to the atmosphere.

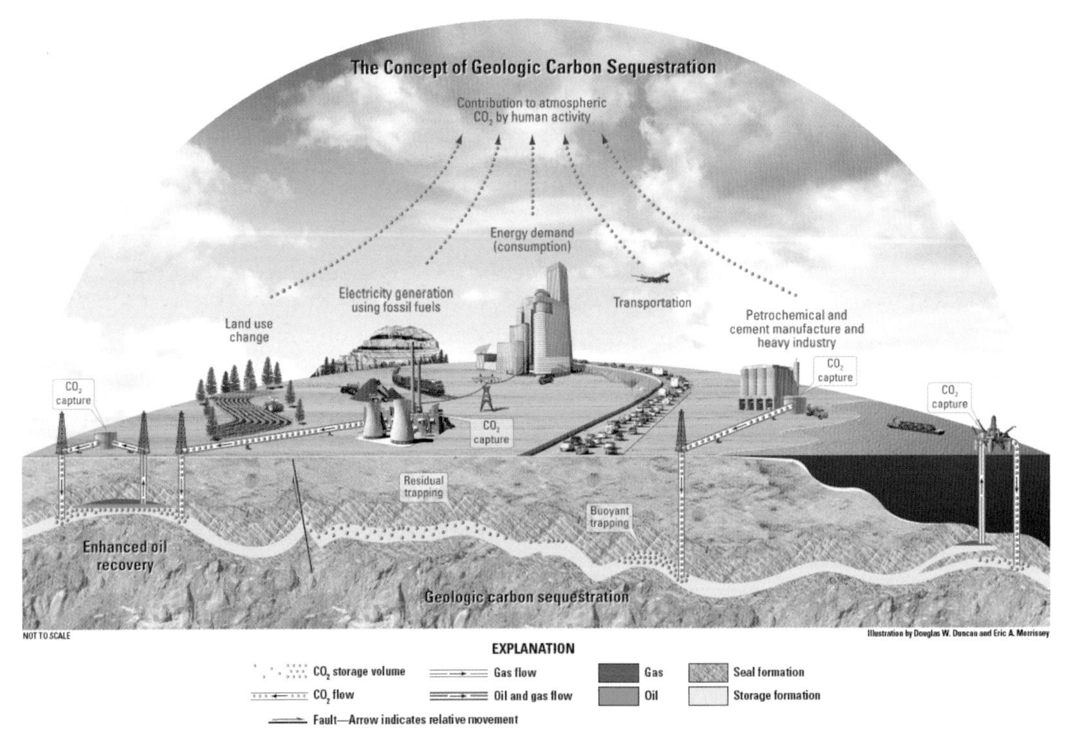

FIGURE 3.14 Illustration of the concept of carbon sequestration. SOURCE: USGS; Duncan and Morrissey (2011).

Technology Background

Geologic formations considered suitable for underground storage of CO_2 include oil and gas reservoirs, unmineable coal seams, and deep saline rock formations (Kaldi et al., 2009). Naturally occurring CO_2 has been trapped in geologic formations for millions of years, which indicates that retaining injected CO_2 in the Earth under the right geological conditions is possible. Injection of CO_2 for EOR has been used in the oil and gas industry for many decades with no obvious adverse effects (see the section Conventional Oil and Gas Production Including Enhanced Oil Recovery, this chapter); CO_2 has also been injected in small volumes into saline rock formations in the western United States and Canada since 1989 without negative consequences (NETL, 2012; Price and Smith, 2008). Saline rock formations used for this purpose are sedimentary rocks that are naturally saturated with highly saline water that is otherwise unsuitable for humans, livestock, or agriculture.

Individual large, coal-fired power plants in the United States produce CO_2 emissions that amount to up to 25 million metric tonnes (~27 million tons) per year.[8] Capturing and transporting CO_2 from industrial plants is technologically possible but is currently expensive, though a significant amount of research is exploring ways to bring costs down (Melzer, 2011). The United States as a whole accounted for approximately 1.5 billion metric tonnes (~1.7 billion tons) of CO_2 emissions in 2010 (EIA, 2012). Storing even a portion of this amount of CO_2 would require capturing the gas at many locations around the country and transporting it to facilities that could inject the CO_2 into appropriate subsurface rock formations.[9]

Efficient underground storage of CO_2 requires that it be in the supercritical (liquid) phase to minimize required storage volume.[10] For CO_2 to remain in a supercritical phase, the confining pressure in the reservoir must be greater than 7.3 MPa (about 73 atm[11]) and temperatures greater than 31.1°C, which can be achieved at depths greater than about 2,600 feet (790 m) (Buruss et al., 2009). These conditions require that the CO_2 be injected at high pressures (62-64 bars [6.2-6.4 MPa or 900-930 psig] at the well head) so that the CO_2 stays as a liquid. The density of supercritical CO_2 is in the range of 0.60-0.75 g/cm^3

[8] See Carbon Monitoring for Action, available at carma.org/.

[9] EOR operations do pump CO_2 underground. However, EOR operations are designed to roughly balance the natural pressure in a reservoir from pumping out of hydrocarbons with pumping in of CO_2. EOR using CO_2 injection currently accounts for approximately 6 percent of U.S. crude oil production (Koottungal, 2010). Natural CO_2 fields are currently the dominant source of CO_2 for U.S. EOR and provide approximately 45 million metric tonnes (~50 million tons) per year, whereas anthropogenic sources, such as CO_2 captured from industrial facilities, account for approximately 10 million metric tonnes (~11 million tons) per year (Kuuskraa, 2010). One of the biggest challenges for EOR projects that wish to use CO_2 injection is being able to secure enough CO_2 consistently at an acceptable cost (Melzer, 2011).

[10] One pound of liquid CO_2, which is about the volume of a typical fire extinguisher, will expand to approximately 8.8 cubic feet (0.25 m^3) at normal room temperature and pressure.

[11] One unit of atmospheric pressure or 1 atm is equivalent to the pressure exerted by the Earth's atmosphere on a point at sea level.

(Sminchak and Gupta, 2003), whereas the density of most formation fluids within potential reservoirs is higher, typically 1.05-1.30 g/cm^3. Supercritical CO_2 is also less viscous than saline formation fluids. These differences in density and viscosity mean that the liquid CO_2 will behave buoyantly within the reservoir. This buoyancy is what makes CO_2 an effective fluid for EOR (Szulczewski et al., 2012).

For CCS, however, the buoyancy of CO_2 means that the geologic reservoir must have a covering of impermeable rock (a "seal") to ensure that the CO_2 will not escape upward (Szulczewski et al., 2012). Depending on the composition of the geologic reservoir for the injected CO_2, some potential exists for supercritical CO_2 either to dissolve, weaken, or transform existing minerals or to precipitate new minerals in the geologic reservoir. For these reasons, selection of a suitable reservoir in which to inject and store CO_2 is critical.

The effects of supercritical CO_2 on geologic materials and the potential impacts of geochemical reactions with brines, cements, casing materials in injection wells, and materials that may seal faults and fractures in the reservoir have been topics of research supported by the Department of Energy (DOE) at academic institutions and national laboratories, and also by the petroleum industry. For example, in 2009 DOE supported 11 projects to conduct site characterization of promising geological formations for CO_2 storage.[12] Research at DOE's National Energy Technology Laboratory (NETL) is based on developing efficient injection techniques, protocols that assess and minimize the impacts of CO_2 on geophysical processes, and remediation technologies to prevent or reduce CO_2 leakage. Currently NETL lists 37 active projects that address the critical geologic barrier for CO_2 storage.[13]

The volumes of supercritical CO_2 discussed for CCS are extremely large. An Intergovernmental Panel on Climate Change special report on CO_2 capture and storage suggests that between approximately 97 and 306 million m^3 per year (converted from 73 and 183 million metric tonnes)[14] of CO_2 could be captured and stored worldwide from coal and a similar amount from natural gas energy plants (Metz et al., 2005). This amount is equivalent to approximately 40,000 to 120,000 Olympic size swimming pools. For comparison, over 300 million m^3 of crude oil were produced in the United States in 2010 (over 4 billion m^3 were produced worldwide) (see Table 3.3). It is anticipated that CCS would take place at a number of locations, ideally places near power plants that produce CO_2 so as to avoid long transportation distances. Many of the facilities would be expected to inject CO_2 volumes on the order of several million tonnes (equivalent to several million cubic meters) or more into the ground each year (e.g., Szulczewski et al., 2012). Globally, only a few small-scale commercial CCS projects (the committee defines small-scale as about

[12] See www.fossil.energy.gov/recovery/projects/site_characterization.html.

[13] See www.netl.doe.gov/technologies/carbon_seq/corerd/storage.html.

[14] As the density of supercritical CO_2 ranges from 600 to 750 kg/m^3, the volume of 1 million metric tonnes (~1.1 million tons) of supercritical CO_2 ranges from 1.33 to 1.67 million m^3. In-ground storage volume will depend on the effective porosity (i.e., the porosity times the storage efficiency).

TABLE 3.3 Petroleum and Natural Gas Production in 2010

	Crude Oil	Natural Gas Plant Liquids (NGPL)	Other Liquids	Total Crude Oil, NGPL, and Other Liquids		Total Dry Natural Gas
United States	2.00 billion barrels	757 million barrels	391 million barrels	3.15 billion barrels	**501 million m^3**	**611 billion m^3**
World	27.0 billion barrels	3.08 billion barrels	754 million barrels	30.9 billion barrels	**4.91 billion m^3**	**3.17 trillion m^3**

NOTE: 1.00000 barrel = 0.15899 m^3.
SOURCE: EIA 2010 International Energy Statistics (available at www.eia.gov/cfapps/ipdbproject/IEDIndex3.cfm).

1 million metric tonnes [approximately 1.55 million m^3][15] or less of CO_2 stored per year in geologic reservoirs) are in operation. In the United States, no commercial CCS technologies are currently deployed, although DOE-supported research is currently exploring the most suitable technologies for CCS through regional partnerships throughout the country.[16] One of these regional projects in Illinois has advanced to the stage of conducting a large-scale test to inject 1 million metric tonnes of CO_2; DOE defines "large-scale" as 1 million metric tonnes [approximately 1.55 million m^3] or more. Both the global, commercial projects and the Illinois test project are discussed in the sections that follow.

Current Projects

The Norwegian state oil company Statoil and its partners currently operate CCS projects at offshore sites in the Sleipner field on the Norwegian continental shelf and in the In Salah gas field in Algeria (Box 3.8). They had also operated a CCS project at the Snøhvit field in the Barents Sea, north of Norway, until early in 2011. At Sleipner approximately 1 million metric tonnes a year have been injected since 1996. The demonstration CO_2 injection project in northern Illinois has been in development for several years; injection of CO_2 began in late 2011. The project plans to inject approximately 1 million metric tonnes per year for several years (Box 3.9). Seismic activity is being routinely monitored at all of these CCS sites. Although the CO_2 injection rates and volumes for these projects are

[15] Volume calculated using 0.70 g/cm^3 as the density of supercritical CO_2; however, this density may range from 0.60 to 0.75 g/cm^3.
[16] See fossil.energy.gov/programs/sequestration/partnerships/index.html.

BOX 3.8
The Sleipner, Snøhvit, and In Salah CO_2 Capture and Storage Projects

In 1996, the Sleipner oil and gas fields in the North Sea became the site of the world's first and largest offshore commercial CO_2 capture and storage project. Carbon dioxide is captured at a plant located on one of the field's operating offshore natural gas platforms and is stored underground in a sandstone formation at depths of approximately 800-1,100 m below the sea bed. Motivation for the project derived from a CO_2 offshore tax levied on offshore oil and gas operations by the Norwegian government in 1991. CO_2 is removed from the natural gas produced at Sleipner and is reinjected into the subsurface into a very porous, permeable sandstone and saline aquifer, the Utsira Formation (Figure 1). The Utsira Formation has an unusually high porosity and permeability (porosity is between 0.35 and 0.4 and permeability is near 1,000 mD) compared to the CO_2 reservoirs in the other two Statoil CCS projects (Figure 2) and to many other potential CCS reservoirs. Approximately 1 million metric tonnes (1.1 million tons) of CO_2 have been stored per year since operations began—with the accumulated total CO_2 in the formation at the middle of 2012 approximately 13.5 million tonnes (Eiken and Ringrose, personal communication). The project is designed for approximately 25 years of CO_2 injection. Current estimates for the Utsira Formation storage capacity range from 2 to 15.7 billion tonnes of CO_2 (NPD, 2011).

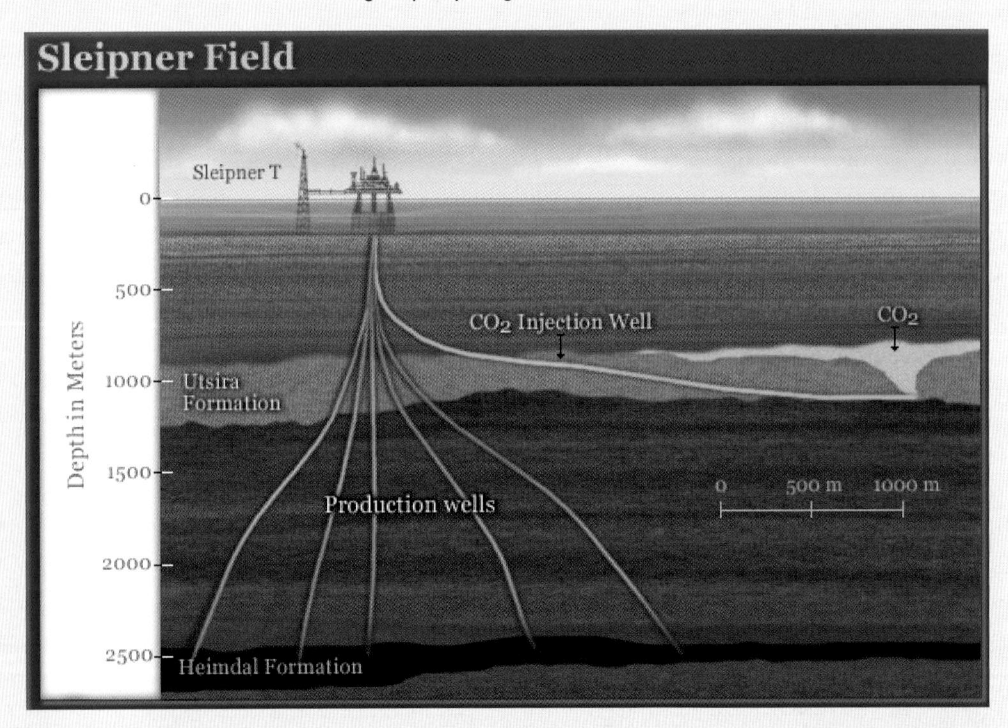

Figure 1 Schematic rendition of the Sleipner field with CO_2 injection into the Utsira sandstone formation occurring as natural gas is extracted from the Heimdal Formation more than 1,000 m below the CO_2 reservoir. SOURCE: © 2012 Schlumberger Excellence in Educational Development, Inc. All rights reserved. Available at www.planetseed.com/node/15252.

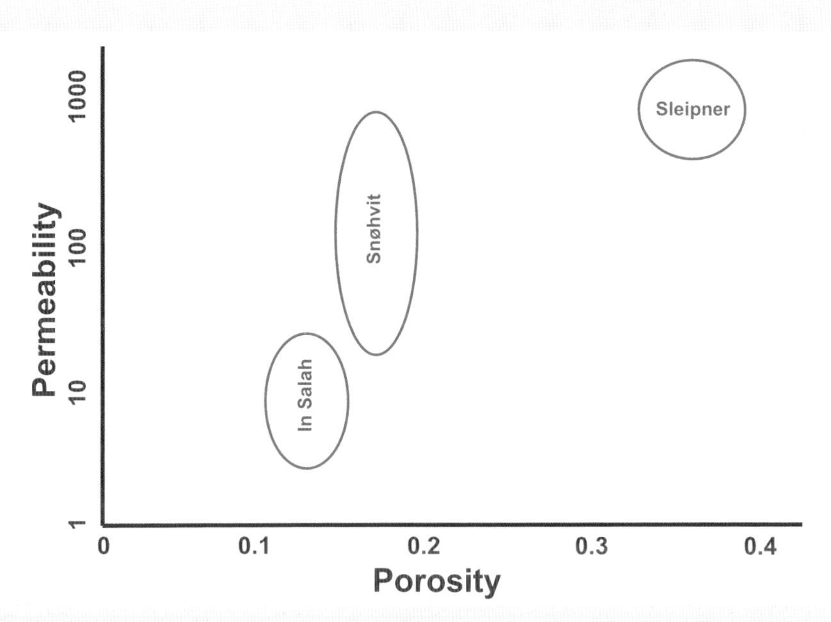

Figure 2 Comparison of porosity and permeability for the CO_2 reservoirs in each of the three projects. The Utsira Formation in the Sleipner field has an unusually high porosity and permeability. SOURCE: Eiken et al. (2011).

The Snøhvit field offshore northern Norway is a natural gas field with an onshore liquid natural gas (LNG) facility. Carbon dioxide separated during the LNG process was captured at the plant and piped back to the field, where it was reinjected underground into a sandstone formation ~2,600 m (8,560 feet) below the seafloor and below the main natural gas reservoir for the gas field; the entire offshore facility is subsea and operated remotely from shore (Statoil, 2009). Carbon storage began in 2008, and CO_2 injection for storage was changed from the Tubåen Formation to the gas-producing Stø Formation in March 2011. Monitoring throughout the injection phase revealed increases in reservoir pressure beyond what had been initially anticipated, indicating that the reservoir had a lower capacity to inject or store CO_2 than had been calculated at the start of the project (Helgesen, 2010). Total stored CO_2 through March 2011 was about 1.1 megatons (Eiken and Ringrose, 2011).

At the In Salah field at Krechba onshore Algeria, the operators began injecting CO_2 in 2004 into a formation located at intermediate depths between Sleipner and Snøhvit. By early 2011, nearly 4 million tons (3.6 million metric tonnes) of CO_2 had been injected. The field has five gas production wells and three CO_2 injection wells. The CO_2 for injection derives from both the produced gas at the field and from gas produced at other fields that is piped to the injection well (Eiken and Ringrose, 2011).

The injection histories for all three fields are shown in Figure 3. The injection at Sleipner was very smooth over 15 years with good injectivity and no evidence of pressure buildup. Very consistent injection pressures were maintained at about 64-65 bars over the course of the project. The conditions at the other two fields proved to be more challenging, with measured pressure increases and limitations on the total capacity of the storage

Box continues

BOX 3.8 Continued

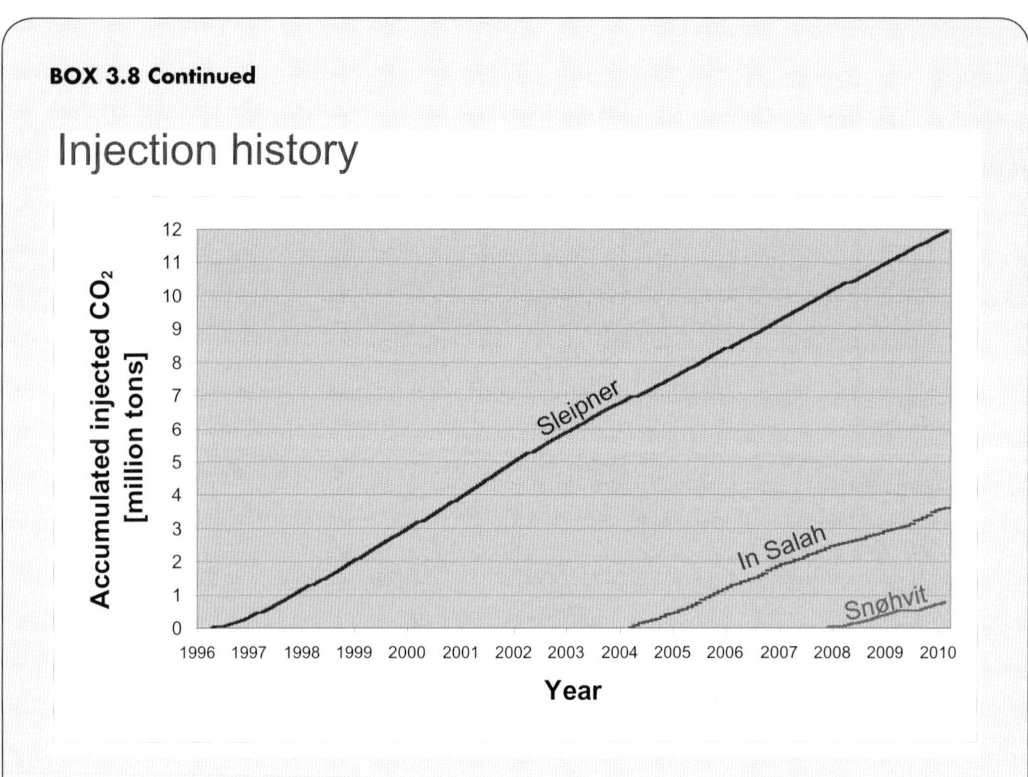

Figure 3 CO_2 injection history at Statoil's Sleipner, Snøhvit, and In Salah fields. SOURCE: Eiken et al. (2011).

formations. Pressure management was deemed an important issue with downhole pressure gauges of great importance (Eiken and Ringrose, 2011).

Prior to the start of all three projects, extensive monitoring was conducted to establish baseline conditions, including any microseismic activity. Monitoring during CO_2 injection for possible leakage and induced seismicity has occurred in all three projects. At both offshore projects, monitoring methods have included measurements of wellhead pressure and temperature, downhole pressure, gravity, and time-lapse seismic. At In Salah, monitoring data have included time-lapse seismic; pressures, rates, and gas chemistry at the wellhead; cores, logs, and fluid samples from the subsurface; one microseismic well, five shallow aquifer wells, and an appraisal well; satellite surveys to measure surface deformation; and surface measurements to monitor for potential leakage or rock strain. Monitoring from pilot wells at this location has shown detectable microseismic events related to CO_2 injection. Shallow wells with three-component seismic detectors are emerging as the preferred deployment solution to give more extensive areal coverage of the field.

SOURCES: Eiken and Ringrose (2011); Eiken and Ringrose (personal communication, June 4, 2012); Ringrose and Eiken (2011); NPD (2011); Helgesen (2010); Statoil (2009); Arts et al. (2008); and "Sleipner Vest" (available at www.statoil.com/en/TechnologyInnovation/ProtectingTheEnvironment/CarboncaptureAndStorage/Pages/CarbonDioxideInjectionSleipnerVest.aspx).

BOX 3.9
Carbon Dioxide Sequestration in the Illinois Basin:
The Midwest Geological Sequestration Consortium Project at Decatur, Illinois

The Midwest Geological Sequestration Consortium (MGSC) is one of seven regional partnerships with funding from the DOE to test methods for geological storage of CO_2. The MGSC in collaboration with Archer Daniels Midland Company, Schlumberger Carbon Services, Trimeric Corporation, and supporting subcontractors has initiated the Illinois Basin-Decatur Project (IBDP), which has begun the injection of 1 million metric tonnes (~1.1 million tons) of supercritical CO_2 over a 3-year period into a saline reservoir that has not had previous fluid extraction at a site near Decatur, Illinois (Figures 1 and 2).

The target reservoir is the Mt. Simon Sandstone, which lies at a depth of approximately 7,000 feet. Injection of CO_2 began in fall 2011 at an initial rate of 1,000 metric tonnes/day. An active seismic surface survey completed in January 2010 prior to the start of injection and a seismic monitoring well are part of the efforts to both monitor the distribution of CO_2 and assess the seismicity risk during injection. The objectives of the baseline survey were to check for faulting, assess reservoir heterogeneity, map reservoir properties, develop data for the mechanical Earth model, and record a baseline for future CO_2 distribution

Box continues

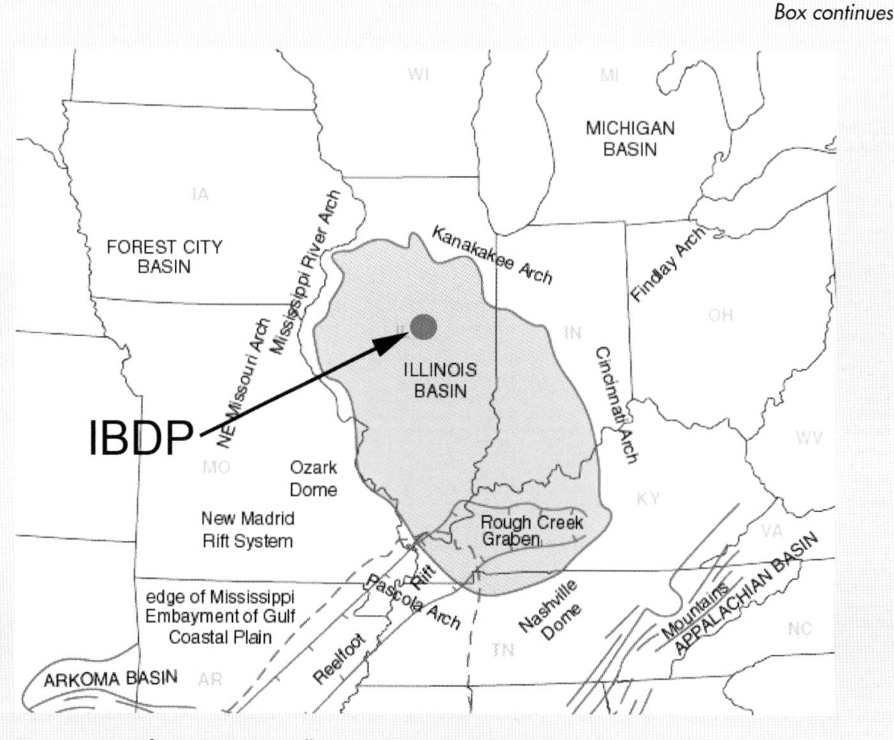

Figure 1 Location of IBDP. SOURCE: Illinois State Geological Survey.

BOX 3.9 Continued

Figure 2 Location of MGSC monitoring well and injection and geophone wells. SOURCE: Illinois Department of Transportation, November 8, 2010.

in the subsurface. Microseismic monitoring is accomplished in both the injection well and a specially drilled microseismic monitoring well. A network for detecting and reporting microseismic events greater than an established magnitude has been installed. The installed array at the IBDP site detected a **M** 3.8 event near Elgin, Illinois, in February 2010, more than a year before the first CO_2 injection. As part of their efforts to develop the CCS projects, the DOE and its collaborators have undertaken a very organized campaign of public outreach and education (see NETL, 2009).

smaller than those being proposed for large power plant and industrial plant operations,[17] these projects provide data for assessment of the potential for induced seismic activity associated with large-scale CCS.

Induced Seismicity Risks

The risk of induced seismicity from CCS is currently difficult to assess accurately. The NETL reported that no harmful induced seismicity had been associated with any of the global CCS storage demonstration projects as of February 2011.[18] However, the volumes of CO_2 injected at these sites so far are small in comparison to the volumes being considered for future proposed large CCS projects. Unlike most water disposal wells, CCS involves continuous CO_2 injection at high rates under high pressures for very long periods of time. The potential therefore exists to increase pore pressures throughout a volume with the storage reservoir that is much larger than those affected by other energy technologies. Given that the potential magnitude of an induced seismic event correlates strongly with the fault rupture area, which in turn relates to the magnitude of pore pressure increase and the volume in which it exists, it would appear that CCS may have the potential for significant seismic risk. The combination of hydro-chemical-mechanical effects such as mineral dissolution may also exacerbate the problem (Espinoza et al., 2011). Some factors could also serve to mitigate risk such as low viscosity and lower injection pressure and limits of permanent pressure change in the reservoir depending upon variables such as reservoir thickness.

DISCUSSION

Geothermal, enhanced geothermal, oil and gas, unconventional oil and gas, and CCS technologies all involve fluid withdrawal and/or injection, thereby providing the potential to induce seismic events. The rates, volumes, pressure, and duration of the injection vary with the technology as do the potential sizes of the earthquakes, the mechanisms to which the earthquakes are attributed (Table 3.4), and the possible risk and hazards of the induced events.

Induced seismicity is commonly characterized by large numbers of small earthquakes that persist during, and in some cases significantly after, fluid injection or removal. At several sites of seismicity caused by or likely related to energy technologies, calculations based on the measured injection pressure and the measured or the inferred state of stress in

[17] Approximately 3,000 million metric tonnes (~3,300 million tons) of CO_2 are reported to have been emitted by the United States in 2009 from the combined activities of electricity and heat production, manufacturing and construction, and other industrial processes including petroleum refining, hydrocarbon extraction, coal mining, and other energy-producing industries. Data available at www.iea.org/co2highlights/co2highlights.pdf.

[18] See www.netl.doe.gov/technologies/carbon_seq/FAQs/permanence4.html.

TABLE 3.4 Summary Information about Historical Felt Induced Seismicity Caused by or Likely Related to[a] Energy Technology Development in the United States

Energy Technology	Number of Projects	Number of Felt Induced Events	Maximum Magnitude of Felt Event	Number of Events **M** ≥ 4.0[b]	Net Reservoir Pressure Change	Mechanism for Induced Seismicity
Vapor-dominated geothermal	1	300-400 per year since 2005	4.6	1 to 3 per year	Attempt to maintain balance	Temperature change between injectate and reservoir
Liquid-dominated geothermal	23	10-40 per year	4.1[c]	Possibly one	Attempt to maintain balance	Pore pressure increase
Enhanced geothermal systems	~8 pilot projects	2-10 per year	2.6	0	Attempt to maintain balance	Pore pressure increase and cooling
Secondary oil and gas recovery (waterflooding)	~108,000 (wells)	One or more felt events at 18 sites across the country	4.9	3	Attempt to maintain balance	Pore pressure increase
Tertiary oil and gas recovery (EOR)	~13,000	None known	None known	0	Attempt to maintain balance	Pore pressure increase (likely mechanism)
Hydraulic fracturing for shale gas production	35,000 wells total	1	2.8	0	Initial positive; then withdraw	Pore pressure increase
Hydrocarbon withdrawal	~6,000 fields	20 sites	6.5	5	Withdrawal	Pore pressure decrease
Wastewater disposal wells	~30,000	8	4.8[d]	7	Addition	Pore pressure increase
Carbon capture and storage, small scale	1	None known	None known	0	Addition	Pore pressure increase
Carbon capture and storage, large scale	0	None	None	0	Addition	Pore pressure increase

TABLE 3.4 Continued

[a]Note that in several cases the causal relationship between the technology and the event was suspected but not confirmed. Determining whether a particular earthquake was caused by human activity is often very difficult. The references for the events in this table and the ways causality may be determined are discussed in the report. **Also important is the fact that the well numbers are those wells in operation today, while the numbers of events listed extend over a total period of decades**.

[b]Although seismic events **M** > 2.0 can be felt by some people in the vicinity of the event, events **M** ≥ 4.0 can be felt by most people and may be accompanied by more significant ground shaking, potentially causing greater public concern.

[c]One event of **M** 4.1 was recorded at Coso, but the committee did not obtain enough information to determine whether or not the event was induced.

[d]**M** 4.8 is a moment magnitude. Earlier studies reported magnitudes up to **M** 5.3 on an unspecified scale; those magnitudes were derived from local instruments.

the Earth's crust suggest that the theoretical threshold for frictional sliding along favorably oriented preexisting fractures was exceeded (see also Chapter 2).

Figure 3.15 shows histograms of the maximum magnitudes reported for induced seismicity associated with different energy technologies: geothermal energy, hydrocarbon extraction, fluid injection for secondary and tertiary oil and gas recovery, hydraulic fracturing associated with unconventional oil and gas production, and wastewater disposal from any of the energy technologies (injection wells) (see Appendix C for data sources for this figure); note that CCS is not included in this figure due to the absence of any known significant induced seismic events associated with this technology.

The largest seismic events and most numerous reports of induced seismicity are associated with extraction activities, with magnitudes up to 7 associated with extraction of gas at the Gazli field. The next largest set of seismic events (two sites in the world, one with an event of **M** 5.1 and another site with an event of **M** 6) is associated with injection activities related to waterflooding for secondary recovery in oil and gas production. Waste and wastewater disposal activities have produced some moderate earthquakes (**M** ~ 4.5), notably in Denver in 1967 at the Rocky Mountain Arsenal, but these are rare. Oklahoma, Colorado, and Arkansas have experienced a recent increase in seismic activity; these events are being examined for potential links to injection (Ellsworth et al., 2012). In the New Mexico–Colorado border area, the Raton Basin is an active coalbed methane field that has experienced several swarms of seismic events, including a **M** 5.3 in August 2010. In light of the seismicity in the Raton Basin, the Colorado Geological Survey (CGS) is now reviewing all permit applications for water disposal wells in Colorado in regard to the possibility of induced seismicity, assisting the Colorado Oil and Gas Conservation Commis-

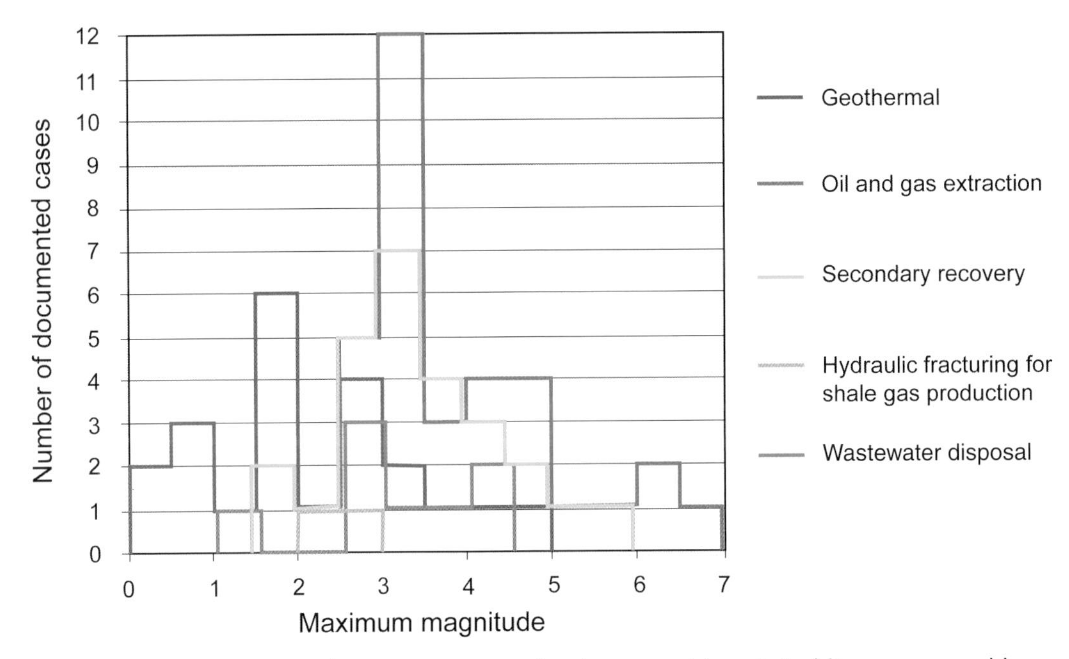

FIGURE 3.15 Histograms of maximum magnitudes documented in technical literature caused by or likely related to subsurface energy production globally. Note: Many gas and oil fields undergo extraction of hydrocarbons along with injection of water for secondary recovery, but if the reported total volume of extracted fluids exceeds that of injection, the site is categorized as extraction. Some cases of induced seismicity in the list above do not have reported magnitudes associated with earthquakes, and those cases are not included in the counts used to develop this figure. No induced seismic events have been recognized related to existing CCS projects. SOURCE: See Appendix C.

sion (COGCC) (CGS, 2012) in the injection well permitting process. The injection and seismicity in the Raton Basin are under close scrutiny by both the CGS and COGCC. A definitive link to injection has not been established in the Raton Basin seismicity. Enhanced seismic arrays have been installed since 2011 in the area and will continue to be studied in detail by field operators, the Colorado agencies, and the USGS.

Numerous geothermal sites report induced seismicity, but the associated maximum magnitudes are generally small, with a maximum reported **M** 4.6 (at The Geysers site in California). Finally, felt seismic events caused by hydraulic fracturing are small and rare, with only one incident globally of hydraulic fracturing causing induced seismicity less than **M** 3 (in Blackpool, England; note the description in Appendix J of the seismic event in Eola, Oklahoma).

Several authors have observed that the maximum magnitudes of seismic events induced by various causes are related to the dimension or volume of human activity. Figure 3.16 (modified from Figure 3 of Nicol et al., 2011) plots the largest earthquake magnitudes

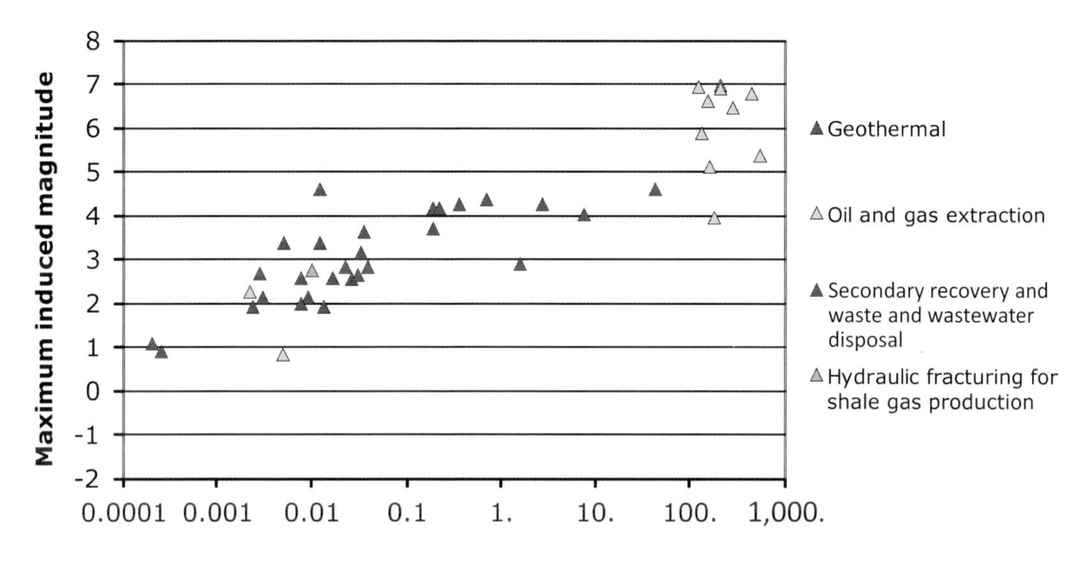

FIGURE 3.16 Graph showing maximum induced seismic event magnitude versus volume of fluid injected into or extracted from single wells or fields that are documented to have had a seismic event directly attributed to or strongly suggested to be caused by one of the energy technologies. These are global data. Events and associated volumes are identified by technology: red triangles denote geothermal energy with most of the data points representing fields (note that the net fluid volume, injected and withdrawn, at The Geysers is actually close to or below zero; see also Figure 3.17); blue triangles denote injection for secondary recovery or waste injection (such as at the Rocky Mountain Arsenal), almost all of which represent single wells; yellow triangles denote fluid extraction (oil or gas withdrawal; note that no data were available on the amount of fluid that may also have been injected in these fields to facilitate withdrawal); and green triangles denote hydraulic fracturing for shale gas production, both of which represent single wells. Not plotted are data from some projects that do not represent maximum magnitude seismic events for that project. Geothermal, extraction, and injection data modified from Figure 3 of Nicol et al. (2011). Hydraulic fracture data have been added in this study.

strongly suggested to be associated with fluid injection or extraction versus the volume of fluid reported for the injection or extraction project. The reported data suggest a correlation between the induced earthquake magnitudes and volumes of fluid injected. McGarr et al. (2002) suggested a correlation between maximum induced magnitude and the scale of human activity by plotting the maximum induced magnitude versus the dimension of the human activity (e.g., the maximum dimension of the hydrocarbon activity). Several points are important regarding these apparent correlations between induced magnitude and fluid volume:

1. Many factors are important in the relationship between human activity and induced seismicity: the depth, rate, and net volume of injected or extracted fluids, bottom-hole pressure, permeability of the relevant geologic layers, locations and properties of faults, and crustal stress conditions. These factors, some of which are interdependent, are also described in Chapter 2. For an induced seismic event to occur, at least two criteria have to be satisfied: (1) the pore pressure change in the reservoir has to exceed a certain critical threshold and (2) a certain net volume of fluid has to be injected (or extracted) to achieve a particular magnitude. The available data suggest, but do not prove, that the net volume of fluid may serve as a proxy for these factors, which indicates what set of conditions will generate small and large earthquakes. Particularly because the other data—bottom-hole pressure, permeability of the relevant geological layers, crustal stress factors, high-resolution well data (full waveform dipole and resistivity and waveform borehole imaging logs), seismic reflection images (two- and three-dimensional surface seismic techniques, 3D vertical seismic profiles or cross well seismic data) to reveal the subsurface structure such as the location, orientation, and properties of faults in the area—are not generally available, total volume can be a tool to draw inferences about various technologies. However, a pure causal relationship between the largest induced magnitudes and fluid volume should not be assumed. Important also, exceptions occur in those cases where fluids are injected into sites such as depleted oil, gas, or geothermal reservoirs, or at sites where the volume of extracted fluids essentially equals or exceeds the volume injected. In those cases pore pressures may not reach the original levels, or in some cases may not increase at all due to the relative volumes of injection and extraction. These data (specifically for oil and gas withdrawal and geothermal energy) are included in Figure 3.16, but it is noted that these specific data points do not necessarily represent the total (net) fluid (injected *and* withdrawn) that may be related to the maximum magnitude event.
2. The volumes indicated in Figure 3.16 include both volumes for individual wells in single projects and volumes for fields. The data cannot be used to predict earthquake magnitudes for an entire region or industry, but rather only to infer what magnitudes might be possible for individual wells or fields.
3. The data in Figure 3.16 are maximum magnitudes associated with fluid injection or extraction and support the requirement, outlined in Chapter 2 and elsewhere in this chapter, that a certain net volume of fluid has to be injected to cause a seismic event of a certain magnitude (or in a similar sense for net fluid withdrawal). The graph does not represent causality, but a condition for an induced seismic event of a certain magnitude to occur. Importantly, the correlation in the figure does not predict what earthquake magnitude will be induced by a specific project, but it reports instead the observed limits (to date) of what earthquake magnitudes have

been observed and can be used to infer what might be the size of the largest induced seismic events, if the volume of injected or extracted fluid is known. However, the correlation cannot be used to directly infer hazard or risk associated with various energy technologies.

4. These data and the limitations described point toward the great value in collecting information about well projects and characteristics, including the size of earthquakes produced (if any). Data are critical to making progress in estimating hazard and risk (see Chapter 5).

Another important factor to consider in evaluating the potential for an energy project to induce felt seismic events is the variation in volume from technology to technology, and the variation in net volume over time (Figure 3.17). For example, although CCS does not have the highest daily injection volumes among the technologies investigated, it does have the highest annual injected volumes because the projects are designed to run continuously with relatively large injection volumes. Also, CCS, similar to waste and wastewater disposal, involves only net addition of fluid to a reservoir rather than both injection and extraction that occur with oil and gas production and geothermal energy development. This characteristic is represented in the bottom graph in Figure 3.17 by the high net volumes of fluid injected for both technologies. Comparatively, the two geothermal cases (The Geysers and the EGS project at Basel) and hydraulic fracturing for shale gas production have negative or low net injection volumes on an annual basis. In the case of The Geysers, the negative net fluid volume is due to the high volumes of fluid extracted; annually, the fluid volume in The Geysers reservoir has actually been declining yearly, despite the high injection volumes.

The tens of thousands of Class II water disposal wells located across the United States have proven to be mostly benign with respect to induced seismicity. However, there are clearly troublesome areas that have induced events as large as **M** 4.7 (Arkansas, 2011; see Horton, 2012) that warrant a closer examination. The dramatic increase in hydraulic fracturing over the past 5 years means an increased volume of wastewater from hydraulic fracturing requiring disposal. If the number of available Class II wastewater disposal wells remains the same, the volume of injected fluid in each well must increase to accommodate the increased wastewater. The long-term effect of this increased volume on the potential to induce felt seismic events is unknown but could be of concern.

The implication for subsurface storage demonstration sites, for instance for CO_2, is that pilot plants that inject small volumes of fluid cannot be expected to represent or bound the induced seismicity that might occur for production plants that will inject much larger volumes. Evaluation of production facilities for large-scale CCS thus requires a complete presentation of the risk of induced seismicity and a comprehensive monitoring plan including bottom-hole pressures and time response to different injection regimes.

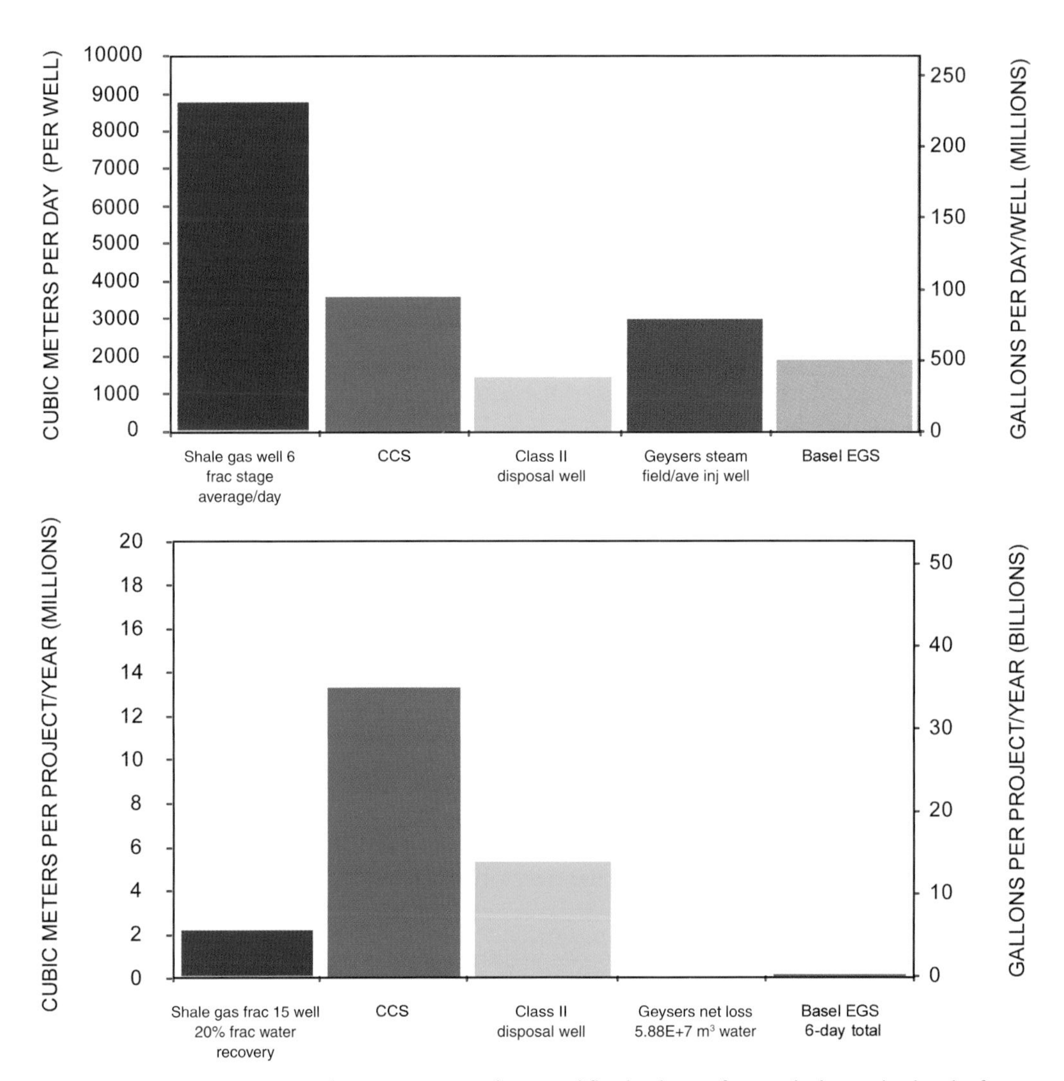

FIGURE 3.17 A comparison showing estimated injected fluid volumes for (1) shale gas hydraulic fracturing, (2) CCS, (3) Class II waste and wastewater disposal wells, (4) The Geysers geothermal steam field for an average injection well, and (5) the Basel EGS project per day (upper graph). The lower graph shows the same information over a 1-year period for each project, with the exception of the Basel EGS project (which operated in total for just 6 days before termination). Data are presented in Appendix L. The committee could not find reliable data per well or per field for hydrocarbon extraction (withdrawal) or for secondary recovery (waterflooding). Hydraulic fracture volumes for shale gas assume a six-stage-per-day program, with a 4.64 million gallon average per well (the "average freshwater volume for fracturing" listed for five shale projects in King, 2012), estimating six hydraulic fracture treatments per day. For the hydraulic yearly volume calculation, an estimate of 15 wells drilled over a project area in the course of a year is made with a 20 percent recovery rate of injected fluid used. The CCS volume shown assumes 1 million tons (~0.9 million metric tonnes) of CO_2 injection per year, similar to the Sleipner field offshore Norway. Class II disposal well data assume 9,000 barrels per day of wastewater injected. The Basel injection volumes averaged 0.5 million gallons per day for 6 days.

REFERENCES

Adushki, V.V., V.N. Rodionov, S. Turuntnev, and A.E. Yodin. 2000. Seismicity in the oil field. *Oilfield Review* Summer:2-17.

Al-Mutairi, S.M., and S.L. Kokal. 2011. EOR Potential in the Middle East: Current and Future Trends. Society of Petroleum Engineers (SPE) conference paper presented at the SPE EUROPEC/EAGE Annual Conference and Exhibition, Vienna, Austria, May 23-26.

Arts, R., A. Chadwick, O. Eiken, S. Thibeau, and S. Nooner. 2008. Ten years' experience of monitoring CO_2 injection in the Utsira Sand at Sleipner, offshore Norway. *First Break* 26:65-72.

Bachmann, C., J. Wössner, and S. Wiemer. 2009. A new probability-based monitoring system for induced seismicity: Insights from the 2006–2007 Basel earthquake sequence. *Seismological Research Letters* 80(2):327.

Barker, B.J., M.S. Gulati, M.A. Bryan, and K.L. Riedel. 1992. Geysers reservoir performance. Pp. 167-177 in *Monograph on the Geothermal Field*, edited by C. Stone, Special Report No. 17, Geothermal Resources Council.

Beall, J.J., M.A. Stark, J.L. Bill Smith, and A. Kirkpatrick. 1999. Microearthquakes in the SE Geysers before and after SEGEP injection. *Geothermal Resources Council Transactions* 23:253-257.

Bhattacharyya, J., and J.M. Lees. 2002. Seismicity and seismic stress in the Coso Range, Coso geothermal field, and Indian Wells valley region, southeast-central California. *Geological Society of America Memoir* 195:243-257.

Blackwell, D.D., and M. Richards. 2004. *Geothermal Map of North America*. Tulsa, OK: American Association of Petroleum Geologists.

Block, L. 2011. Paradox Valley Deep Disposal Well and Induced Seismicity. Presentation to the National Research Council Committee on Induced Seismicity Potential in Energy Technologies, Dallas, TX, September 14.

Boden, T., and T.J. Blasing. 2011. Record high 2010 global carbon dioxide emissions from fossil-fuel combustion and cement manufacture posted on CDIAC site. Carbon Dioxide Information Analysis Center. Available at cdiac.ornl.gov/trends/emis/prelim_2009_2010_estimates.html (accessed April 2012).

Buruss, R.C., S.T. Brennan, P.A. Freeman, M.D. Merrill, L.F. Ruppert, M.F. Becker, W.N. Herkelrath, Y.K. Kharaka, C.E. Neuzil, S.M. Swanson, T.A. Cook, T.R. Klett, P.H. Nelson, and C.J. Schenk. 2009. Development of a probabilistic assessment methodology for evaluation of carbon dioxide storage. U.S. Geological Survey Open-File Report 2009-1035, 81 p. Available at http://pubs.usgs.gov/of/2009/1035/ (accessed May 2012).

CDOGGR (California Division of Oil, Gas and Geothermal Resources). 2011. Oil, Gas & Geothermal—Geothermal Resources. State of California Department of Conservation. Available at www.conservation.ca.gov/dog/geothermal (accessed February 2012).

CGS (Colorado Geological Survey). 2012. Earthquakes Triggered by Humans in Colorado—a background paper by the Colorado Geological Survey. http://geosurvey.state.co.us/hazards/Earthquakes/Documents/Earthquakes%20Triggered.pdf (accessed April 2012).

Cipolla, C.L., and C.A. Wright. 2002 (February). Diagnostic techniques to understand hydraulic fracturing: What? Why? and How? *SPE Production & Facilities* 17(1):23-35.

Clark, C.E., and J.A. Veil. 2009. Produced Water Volumes and Management Practices in the United States. ANL/EVS/R-09/1. Prepared by the Environmental Science Division, Argonne National Laboratory, for the U.S. Department of Energy, Office of Fossil Energy, National Energy Technology Laboratory, September.

Davis, S.D., and W.D. Pennington. 1989. Induced seismic deformation in the Cogdell oil field of West Texas. *Bulletin of the Seismological Society of America* 79(5):1477-1494.

De Pater, C.J., and S. Baisch. 2011. Geomechanical Study of Bowland Shale Seismicity—Synthesis Report. Prepared for Cuadrilla Resources. November 2. Available at http://www.cuadrillaresources.com/wp-content/uploads/2012/02/Geomechanical-Study-of-Bowland-Shale-Seismicity_02-11-11.pdf (accessed June 2012).

Deichmann, N., and D. Giardini. 2009. Earthquakes induced by the stimulation of an enhanced geothermal system below Basel (Switzerland). *Seismological Research Letters* 80(5):784-798.

DOE (U.S. Department of Energy). 2009. Modern Shale Gas Development in the United States: A Primer. Prepared by GWPC and ALL for the U.S. Department of Energy, National Energy Technology Laboratory, 116 pp. Available at http://www.netl.doe.gov/technologies/oil-gas/publications/epreports/shale_gas_primer_2009.pdf (accessed January 2012).

DOE. 2011. Enhanced Oil Recovery/CO_2 Injection. DOE Fossil Energy Office. Available at fossil.energy.gov/programs/oilgas/eor/ (accessed March 2012).

Duncan, D.W., and E.A. Morrissey. 2011. The concept of geologic carbon sequestration. U.S. Geological Survey Fact Sheet 2010-3122. Available at http://pubs.usgs.gov/fs/2010/3122/ (accessed June 2012).

EIA (Energy Information Administration). 1993. Drilling Sideways—A Review of Horizontal Well Technology and Its Domestic Application. DOE/EIA-TR-0565. Available at ftp://tonto.eia.doe.gov/petroleum/tr0565.pdf (accessed January 2012).

EIA. 2011. Review of Emerging Resources: U.S. Shale Gas and Oil Shale Plays. Available at ftp://ftp.eia.doe.gov/natgas/usshaleplays.pdf (accessed November 2011).

EIA. 2012. AEO2012 Early Release Overview. Available at www.eia.gov/forecasts/aeo/er/early_carbonemiss.cfm (accessed February 2012).

Eiken, O., and P. Ringrose. 2011. Statoil experiences on storage and monitoring of CO_2. Presentation (via teleconference) to the National Research Council Committee on Induced Seismicity Potential in Energy Technologies, Irvine, CA, August 18.

Eiken, O., P. Ringrose, C. Hermanrud, B. Nazarian, T.A. Torp, and L. Høier. 2011. Lessons learned from 14 years of CCS operations: Sleipner, in Salah and Snøhvit. *Energy Procedia* 4:5541-5548.

Eisner, L. 2011. Seismicity of DFW, Texas, USA. Presentation to the National Research Council Committee on Induced Seismicity Potential in Energy Technologies, Dallas, TX, September 14.

Ellsworth, W.L., S.H. Hickman, A.L. Llenos, A. McGarr, A.J. Michael, and J.L. Rubinstein. 2012. Are seismicity rate changes in the midcontinent natural or manmade? 2012 Seismological Society of America Annual Meeting abstracts.

Engelder, T. 1993. *Stress Regimes in the Lithosphere*. Princeton, NJ: Princeton University Press, 451 pp.

EPA (Environmental Protection Agency). 2011. Draft Plan to Study the Potential Impacts of Hydraulic Fracturing on Drinking Water Resources. EPA/600/E-11/001/February 2011/www.epa.gov/research. Available at water.epa.gov/type/groundwater/uic/class2/hydraulicfracturing/upload/hf_study_plan_110211_final_508.pdf (accessed January 2012).

Espinoza, D.N., S.H. Kim, and J.C. Santamarina. 2011. CO_2 geological storage—geotechnical implications. *KSCE Journal of Civil Engineering* 15(4):707-719.

Feng, Q., and J.M. Lees. 1998. Microseismicity, stress, and fracture in the Coso geothermal field, California. *Tectonophysics* 289:221-238.

Fialko, Y., and M. Simons. 2000. Deformation and seismicity in the Coso geothermal area, Inyo County, California: Observations and modeling using satellite radar interferometry. *Journal of Geophysical Research* 105:21,781-21,794.

Fisher, K. 2010. Data Confirm Safety of Well Fracturing. *The American Oil and Gas Reporter*. July 10. Available at www.fidelityepco.com/Documents/OilGasRept_072010.pdf (accessed December 2011).

Fisher, K., and N. Warpinski. 2011. Hydraulic Fracture Height-Growth: Real Data. SPE 145949 presented to the Society of Petroleum Engineers (SPE) Annual Technical Conference and Exhibition, Denver, CO, October 30-November 2.

Friedlingstein, P., R.A. Houghton, G. Marland, J. Hacker, and T.A. Boden. 2010. Update on CO_2 emissions. *Nature Geoscience* 3:811-812.

Frohlich, C., E. Potter, C. Hayward, and B. Stump. 2010. Dallas-Fort Worth earthquakes coincident with activity associated with natural gas production. *The Leading Edge* 29(3):270-275.

Grasso, J.-R. 1992. Mechanics of seismic instabilities induced by the recovery of hydrocarbons. *Pure and Applied Geophysics* 139(3/4):506-534.

Hamilton, R.M., and L.J.P. Muffler. 1972. Microearthquakes at The Geysers geothermal area, California. *Journal of Geophysical Research* 77:2081-2086.

Häring, M.O., U. Schanz, F. Ladner, and B.C. Dyer. 2008. Characterization of the Basel 1 enhanced geothermal system. *Geothermics* 37:469-495.

Harper, T.R. 2011. Well Preese Hall-1: The Mechanism of Induced Seismicity. Report prepared by Geosphere Ltd. for Cuadrilla Resources Ltd. Lancashire, UK, October 10.

Helgesen, O.K. 2010. CO_2-problemer på Snøhvit. *TekniskUkeblad*. March 18. Available at www.tu.no/olje-gass/article240108.ece (accessed September 2011).

Holland, A. 2011. Examination of Possibly Induced Seismicity from Hydraulic Fracturing in the Eola Field, Garvin County, Oklahoma. Oklahoma Geological Survey Open-File Report OF1-2011. Available at www.ogs.ou.edu/pubsscanned/openfile/OF1_2011.pdf (accessed February 2012).

Horton, S. 2012. Disposal of hydrofracking-waste fluid by injection into subsurface aquifers triggers earthquake swarm in central Arkansas with potential for damaging earthquake. *Seismological Research Letters* 83(2):250-260.

Kaldi, J.G., C.M. Gibson-Poole, and T.H.D. Payenberg. 2009. Geological input to selection and evaluation of CO_2 geosequestration sites. Pp. 5-16 in *The American Association of Petroleum Geologists Special Volumes*, 1st ed. Tulsa, OK: American Association of Petroleum Geologists.

Kaven, J.O., S.H. Hickman, and N.C. Davatzes. 2011. Micro-seismicity, fault structure and hydraulic compartmentalization within the Coso geothermal field, California. Proceedings of the Thirty-Sixth Workshop on Geothermal Reservoir Engineering, Stanford University, CA, January 31-February 2.

King, G.E. 2010. Thirty years of gas shale fracturing: What have we learned? Presented at the Society of Petroleum Engineers Annual Technical Conference and Exhibition, Florence, Italy, September 19-22.

King, G.E. 2012. Hydraulic fracturing 101: What every representative, environmentalist, regulator, reporter, investor, university researcher, neighbor, and engineer should know about estimating frac risk and improving frac performance in unconventional gas and oil wells. SPE 152596 presented to the Society of Petroleum Engineers (SPE) Hydraulic Fracturing Technology Conference, The Woodlands, TX, February 6-8.

Koenig, J.B. 1992. History of development at The Geysers geothermal field, California. Pp. 7-18 in *Monograph on The Geothermal Field*, Special Report No. 17, edited by C. Stone. Geothermal Resources Council.

Koottungal, L. 2010. Special report: 2010 worldwide EOR survey. *Oil and Gas Journal* 108(14):41-53.

Kraft, T., P.M. Mai, S. Wiemer, N. Deichmann, J. Ripperger, P. Kastli, C. Bachmann, D. Fah, J. Wossner, and D. Giardini. 2009. Enhanced geothermal systems: Mitigating risk in urban areas. *Eos* 90(32):273-280.

Kuuskraa, V.A. 2010. Challenges of implementing large-scale CO_2 enhanced oil recovery with CO_2 capture and storage. White paper prepared for the Symposium on the Role of Enhanced Oil Recovery in Accelerating the Deployment of Carbon Capture and Storage, Cambridge, MA, July 23. Available at web.mit.edu/mitei/docs/reports/eor-css/kuuskraa.pdf (accessed April 2012).

Luza, K.V., and J.E. Lawson, Jr. 1980. Seismicity and tectonic relationships of the Nernaha Uplift in Oklahoma. U.S. Nuclear Regulatory Commission NUREG/CR-1500, Part 3, 70 pp.

Majer, E.L., R. Baria, M. Stark, S. Oates, J. Bonner, B. Smith, and H. Asanuma 2007. Induced seismicity associated with enhanced geothermal systems. *Geothermics* 36:185-222.

McGarr, A., D. Simpson, and L. Seeber. 2002. Case histories of induced and triggered seismicity. Pp. 647-661 in *International Handbook of Earthquake and Engineering Seismology, Part A*, edited by W.H.K. Lee et al. New York: Academic Press.

Melzer, S. 2011. CO_2 Enhanced Oil Recovery. Presentation to the National Research Council Committee on Induced Seismicity Potential in Energy Technologies, Dallas, TX, September 14.

Metz, B., O. Davidson, H. de Coninck, M. Loos, and L. Meyer (eds). 2005. *Carbon Dioxide Capture and Storage*. Intergovernmental Panel on Climate Change (IPCC) Special Report. Cambridge, UK: Cambridge University Press. 431 pp.

Meyer, J.P. 2007. Summary of Carbon Dioxide Enhanced Oil Recovery (CO_2 EOR) Injection Well Technology. Prepared for the American Petroleum Institute. Plano, TX: Contek Solutions. Available at www.gwpc.org/e-library/documents/co2/APIpercent20CO2percent20Report.pdf (accessed February 2012).

MIT (Massachusetts Institute of Technology). 2006. *The Future of Geothermal Energy: Impact of Enhanced Geothermal Systems (EGS) on the United States in the 21st Century*. Boston: MIT Press.

Montgomery, C.T., and M.B. Smith. 2010. Hydraulic fracturing: History of an enduring technology. *Journal of Petroluem Technology* December:26-32. Available at http://www.spe.org/jpt/print/archives/2010/12/10Hydraulic.pdf (accessed January 2012).

NETL (National Energy Technology Laboratory). 2007. DOE's Unconventional Gas Research Programs 1976-1995. U.S. Department of Energy. Available at http://www.netl.doe.gov/kmd/cds/disk7/disk2/Final%20Report.pdf (accessed December 2011).

NETL. 2009. Best Practices for: Public Outreach and Education for Carbon Storage Projects. DOE/NETL-2009/1391. U.S. Department of Energy. Available at www.netl.doe.gov/technologies/carbon_seq/refshelf/BPM_PublicOutreach.pdf (accessed December 2011).

NETL. 2010. Carbon Dioxide Enhanced Oil Recovery: Untapped Domestic Energy Supply and Long Term Carbon Storage Solution. U.S. Department of Energy. Available at www.netl.doe.gov/technologies/oil-gas/publications/EP/CO2_EOR_Primer.pdf (accessed January 2012).

NETL. 2012. Carbon Sequestration FAQ Information Portal: Permanence and Safety of CCS. Available at www.netl.doe.gov/technologies/carbon_seq/FAQs/permanence2.html (accessed February 2012).

NRC (National Research Council). 1996. *Rock Fractures and Fluid Flow: Contemporary Understanding and Applications*. Washington, DC: National Academy Press.

NRC. 2001. *Energy Research at DOE: Was It Worth It?* Washington, DC: National Academy Press.

NRC. 2010. *Management and Effects of Coalbed Methane Produced Water in the United States*. Washington, DC: The National Academies Press.

NRC. 2011. *America's Climate Choices*. Washington, DC: The National Academies Press.

NRC. 2012. *Water Reuse: Potential for Expanding the Nation's Water Supply through Reuse of Municipal Wastewater*. Washington, DC: The National Academies Press.

Nicholson, C., and R.L. Wesson. 1990. Earthquake hazard associated with deep well injection: A report to the U.S. Environmental Protection Agency. U.S. Geological Survey Bulletin 1951, 74 pp.

Nicol, A., R. Carne, M. Gerstenberger, and A. Christopher. 2011. Induced seismicity and its implications for CO_2 storage. *Energy Procedia* 4:3699-3706.

Nicot, P., and B.R. Scanlon. 2012. Water use for shale-gas production in Texas, U.S. *Environmental Science and Technology* 46:3580-3586.

NPD (Norwegian Petroleum Directorate). 2011. CO_2 Storage Atlas Norwegian North Sea. Available at http://www.npd.no/Global/Norsk/3-Publikasjoner/Rapporter/PDF/CO2-ATLAS-lav.pdf (accessed June 2012).

Preiss, J.S., S.R. Walter, and D.H. Oppenheimer. 1996. Seismicity maps of the Santa Rosa 1deg by 2deg Quadrangle, California for the period 1969-1995. U.S. Geological Survey Miscellaneous Investigations Series.

Price, J., and B. Smith. 2008. *Geologic Storage of Carbon Dioxide: Staying Safely Underground*. Booklet of the International Energy Agency (IEA) Working Party on Fossil Fuels. Available at www.co2crc.com.au/dls/external/geostoragesafe-IEA.pdf (accessed January 2012).

Ringrose, P. and O. Eiken. 2011. Microseismic monitoring for CO_2 storage. Presentation (via teleconference) to the National Research Council Committee on Induced Seismicity Potential in Energy Technologies, Irvine, CA, August 18.

Shepherd, M. 2009. Oil field production geology. AAPG Memoir 91. American Association of Petroleum Geologists, Tulsa, OK, 350 pp.

Simpson, D.W., and W. Leith. 1985. The 1976 and 1984 Gazli, USSR, earthquakes—Were they induced? *Bulletin of the Seismological Society of America* 75:1465-1468.

Sminchak, J., and N. Gupta. 2003. Aspects of induced seismic activity and deep-well sequestration of carbon dioxide. *Environmental Geosciences* 10(2):81-89.

Smith, J.L.B., J.J. Beall, and M.A. Stark. 2000. Induced seismicity in the SE Geysers Field, California, USA. Proceedings of the World Geothermal Congress, Kyushu-Tohoku, Japan, May 28-June 10.

Soeder, D.J., and W.M. Kappel. 2009. Water Resources and Natural Gas Production from the Marcellus Shale. U.S. Geological Survey Fact Sheet 2009-3032, 6 pp. Available at http://pubs.usgs.gov/fs/2009/3032/ (accessed January 2012).

Statoil. 2009. Carbon Dioxide Capture, Transport and Storage (CCS). Available at www04.abb.com/GLOBAL/SEITP/seitp202.nsf/c71c66c1f02e6575c125711f004660e6/9ebcbfaedaedfa36c12576f1004a49de/$FILE/StatoilHydro+CCS.pdf (accessed October 2011).

Szulczewski, M.L., C.W. MacMinn, H.J. Herzog, and R. Juanes. 2012. Lifetime of carbon capture and storage as a climate-change mitigation technology. *Proceedings of the National Academy of Sciences*. Available at www.pnas.org/content/early/2012/03/15/1115347109 (accessed April 2012).

Texas RRC (Railroad Commission). 2010. Saltwater Disposal Wells Frequently Asked Questions (FAQs). Updated February 1, 2010. Available at www.rrc.state.tx.us/about/faqs/saltwaterwells.php (accessed February 2012).

Williams, C.F., M.J. Reed, R.H. Mariner, J. DeAngelo, and S.P. Galanis, Jr. 2008. Assessment of Moderate- and High Temperature Geothermal Resources of the United States. USGS Fact Sheet 2008-3082.

CHAPTER FOUR

Governmental Roles and Responsibilities Related to Underground Injection and Induced Seismicity

Chapter 3 reviewed several instances of seismic activity that may have been induced by underground injection. Underground injection of fluids is a key component of enhanced oil recovery, development of some unconventional oil and gas resources such as shale gas, geothermal energy production, carbon capture and storage, and wastewater disposal, which is often a part of different kinds of energy technology development. Although seismic events induced by the underground injection of fluids have been recognized for many decades, few of these events have captured national attention. However, the recent debate concerning hydraulic fracturing has brought the issue of induced seismicity to a higher level of public attention. The Environmental Protection Agency (EPA) is studying this topic[1] concurrently with this National Research Council study and will publish its own report on this issue. It is important to note that, although this chapter deals mainly with induced seismicity caused by or likely related to the underground injection of fluid, induced seismicity can also be caused by the withdrawal of fluid from underground geologic formations.

Four federal agencies—the EPA, the Bureau of Land Management (BLM), the U.S. Forest Service (USFS), and the U.S. Geological Survey (USGS)—and different state agencies have regulatory oversight, research roles, and/or responsibilities related to different parts of the underground injection activities that are associated with energy technologies. Understanding these roles and responsibilities is important to the future development of energy technologies in ways that preserve public safety while allowing development of energy resources. This chapter provides a brief description of each agency's authority related to underground injection and induced seismicity. States' roles and responsibilities are also discussed; however, the committee did not perform a comprehensive review of all the states that are active in addressing the issue.

[1] EPA has been facilitating a National Technical Working Group on Injection Induced Seismicity since mid-2011 and anticipates releasing a report that will contain technical recommendations directed toward injection-induced seismicity specific to Underground Injection Control (UIC) and Class II wells. See http://www.gwpc.org/meetings/uic/2012/proceedings/09McKenzie_Susie.pdf; P. Dellinger, presentation to the committee, September 2011.

FEDERAL AUTHORITIES

Environmental Protection Agency

More than 700,000 different wells are currently used for the underground injection of fluids in the United States and its territories.[2] Underground fluid injection began in the 1930s in order to increase production from existing oil and gas fields and was used in later years to dispose of industrial waste, but it was unregulated until 1974 when Congress passed the Safe Drinking Water Act (SDWA). The SDWA ensures safe drinking water for the public and establishes regulatory authority over the underground injection of fluids. In accordance with the act, the EPA is required to set standards for drinking water quality and to oversee all states, localities, and water suppliers that implement these standards. The EPA also regulates the construction, operation, permitting, and final plugging and abandonment of injection wells that place fluids underground for storage or for disposal under its Underground Injection Control (UIC) program.[3] It is important to note that the SDWA gives authority to the EPA to protect underground sources of drinking water from contamination due to underground injection and does not explicitly address the issue of seismicity induced by underground injection. UIC regulations requiring information on locating and describing faults in the area of a proposed disposal well are concerned with containment of the injected fluid, not the possibility of induced seismicity.

Developers applying for a permit to inject fluids underground must demonstrate to the EPA that the operation will not endanger any underground sources of drinking water (USDWs). This regulatory scheme allows for six classes of injection wells, which are classified by the type of fluid injected and the specific injection depth (e.g., above or below sources of drinking water). Under this program, oil and gas industry injection wells are regulated as Class II injection wells, which also generally cover enhanced oil recovery projects or projects involving the disposal of exploration and production wastes (NRC, 2010). Table 4.1 provides an explanation of the distinction among classes of wells regulated under the SDWA.

Although the number and distribution of the different classes of injection wells vary by state, Class V wells are by far the most numerous, accounting for almost 79 percent of the total number of reported UIC wells. Because Class V wells normally inject fluid into formations above USDWs, these wells are usually too shallow to be considered a source of induced seismicity. This does not hold true in all cases, however, because wells used for fluid injection associated with the extraction of geothermal energy are included in this class of injection wells and are often the source of seismic events. The total number of geothermal wells in the United States was estimated to be approximately 239 wells, with 153 of these wells located in California and 53 located in Nevada (EPA, 1999). Although Class VI wells

[2] See water.epa.gov/type/groundwater/uic/basicinformation.cfm.

[3] Ibid.

TABLE 4.1 Classes of Wells in the EPA UIC Program

Class	Use
I	Injection of hazardous wastes, industrial nonhazardous liquids, or municipal wastewater beneath the lowermost underground sources of drinking water (USDWs) (650 wells).
II	Injection of brines and other fluids associated with oil and gas production and hydrocarbons for storage. Injected beneath the lowermost USDW (151,000 wells).[a]
III	Injection of fluids associated with solution mining of minerals beneath the lowermost USDW (21,000 wells).
IV	Injection of hazardous or radioactive wastes into or above USDWs. Banned wells unless authorized by federal or state groundwater remediation project (24 sites).
V	All injection wells not included in Classes I–IV. Generally used to inject nonhazardous fluids into or above USDWs and typically shallow onsite disposal systems (estimated 400,000–650,000 wells).[b]
VI	Inject carbon dioxide (CO_2) for long-term storage, also known as geologic sequestration of CO_2 (estimated 6–10 wells by 2016).

[a] The table provided by EPA describes Class II wells as "injected below the lowermost USDW." Although this is correct in most cases, injection below the lowermost USDW is not required for Class II wells, according to UIC regulations.

[b] Most Class V wells are unsophisticated shallow disposal systems that include storm water drainage wells, cesspools, and septic system leach fields.

SOURCES: water.epa.gov/type/groundwater/uic/wells.cfm and water.epa.gov/type/groundwater/uic/class5/classv_study.cfm.

also inject into formations below USDWs, no commercial carbon sequestration facilities are operating at this time.

Texas, California, and Kansas have the highest number of deep injection wells[4] (counting only Classes I through IV), and 15 states have no deep injection wells at all. Table 4.2 shows the number of UIC wells in each state, listed by well count.

As Table 4.2 shows, Class II injection wells represent 87 percent of the total number of Class I through Class IV wells. For this reason the oil-producing states of Texas, California, Kansas, Wyoming, and Oklahoma have higher numbers of deep injection wells than other states.

States, territories, and tribes can submit an application to the EPA to obtain primary enforcement responsibility, or "primacy," to implement the UIC program within their borders.[5] Agencies that have been granted this authority oversee the injection activities

[4] A deep injection well is a well that injects fluid below all underground sources of drinking water.

[5] See water.epa.gov/type/groundwater/uic/Primacy.cfm.

TABLE 4.2 2010 UIC Well Inventory Sorted by the Total Number of Deep Underground Injection Wells

State	Class 1 Wells	Class II Wells	Class III Wells	Class IV Sites	Total UIC Wells Class I through Class IV
TX	108	52016	6075	4	58203
CA	45	29505	212	0	29762
KS	53	16658	145	0	16856
WY	41	4978	10552	0	15571
OK	6	10629	2	2	10639
IL	5	7843	0	0	7848
NM	5	4585	10	0	4600
NE	3	661	3913	0	4577
LA	37	3731	89	0	3857
KY	2	3403	0	0	3405
OH	10	2455	54	0	2519
IN	28	2091	0	0	2119
PA	0	1861	0	0	1861
MI	30	1460	46	0	1536
AK	29	1347	0	0	1376
MS	5	1110	0	0	1115
AR	13	1093	0	0	1106
MT	0	1062	0	0	1062
ND	4	1023	1	0	1028
CO	13	874	34	0	921
WV	0	779	21	0	800
NY	1	532	174	0	707
UT	0	428	16	8	452
MO	0	282	0	0	282
FL	212	58	0	0	270
AL	0	240	3	0	243
SD	0	87	0	0	87
VA	0	11	6	0	17
TN	0	18	0	0	18
NV	0	18	0	0	18
OR	0	9	0	7	16
AZ	0	0	15	0	15
NC	0	0	0	3	3
IA	0	3	0	0	3
WA	0	1	0	0	1

NOTE: Because fluid in Class I through Class IV wells are normally injected into formations below USDWs, these wells can be a cause of induced seismicity. Class V wells normally inject fluid above USDWs and are normally too shallow to create induced seismicity and are therefore excluded from this table. The 15 states not listed here have no deep injection wells. No Class VI wells are currently in operation; however, 6-10 are estimated by 2016.
SOURCE: EPA (2010).

TABLE 4.3 Status of EPA Regulatory Authority Across the United States

State Program	Alabama, Arkansas, Connecticut, Delaware, District of Columbia, Georgia, Guam, Idaho, Illinois, Kansas, Louisiana, Maine, Maryland, Massachusetts, Mississippi, Missouri, Nebraska, Nevada, New Hampshire, New Jersey, New Mexico, North Carolina, North Dakota, Ohio, Oklahoma, Oregon, Puerto Rico, Rhode Island, South Carolina, Texas, Utah, Vermont, Washington, West Virginia, Wisconsin, Wyoming
EPA Program	American Samoa, Arizona, Hawaii, Iowa, Kentucky, Michigan, Minnesota, New York, Pennsylvania, Tennessee, Virgin Islands, Virginia
State/EPA Program	Alaska, California, Colorado, Florida, Indiana, Montana, South Dakota
Tribal/EPA Program	Fort Peck Tribe, Navajo Nation

SOURCE: EPA; available at water.epa.gov/type/groundwater/uic/Primacy.cfm.

within their state. The EPA remains responsible for issuing permits in states that have not been delegated primacy and for the UIC programs on most tribal lands. Primacy for all classes of injection wells does not need to be granted to a state in order for a state to exercise regulatory authority over a single class of wells. For example, a state may exercise primacy over only Class II wells and no other class of injection wells. In this case the EPA would retain jurisdiction over all other well classes within the UIC program except Class II wells where primacy was delegated to the state. Currently, the EPA has granted primacy over all classes of injection wells in 33 states and 2 territories. The EPA shares jurisdiction for injection regulation in 7 states and has complete regulatory authority over underground injection in 10 states and 2 territories (see Table 4.3).

Primacy allows states to permit facilities, inspect wells, enforce against violations, and otherwise regulate underground injection activity within the state. States with primacy can disperse this authority through different state agencies. Some states regulate all classes of injection wells through one state agency (e.g., the Department of Health and Environment), and others divide the regulatory authority between several state agencies such as oil and gas commissions, health departments, and the local divisions of mining. However, regardless of how jurisdiction is divided, all state regulatory agencies are required to establish regulations that, at a minimum, conform to the EPA's UIC guidelines, which are outlined in Title 40 of the *Code of Federal Regulations* (CFR), Part 145.[6]

The authority delegated to the EPA by the SDWA is limited to technical issues involving well bore construction, allowable sources of injected fluid, and operational requirements such as maximum pressures and periodic testing that protect underground sources of drink-

[6] Available at www.access.gpo.gov/nara/cfr/waisidx_02/40cfr145_02.html.

ing water and the surface environment. The EPA, however, does not grant a contractual right to inject fluids or CO_2 underground by their permitting process. In the case of fluid disposal or CO_2 sequestration, this right is granted by the property owner via a "surface use agreement" with the injection well operator. These agreements may include fees paid to the property owner based on a monetary charge per barrel of fluid or ton of CO_2 or a charge for land rental per month. These agreements can also include requirements on how fluid is delivered (by truck or by pipeline), how site security is handled, and what type of facilities will be used on the well site (tank, pits, and offloading facilities). Property owners can be private parties and/or governmental agencies such the BLM, the USFS, or state land management organizations. Underground injection for the purpose of secondary or tertiary recovery operations in an existing oil or gas field or injection to develop geothermal resources are usually allowed via an oil and gas or geothermal mineral lease.

Specific regulations governing the requirements of the UIC program are documented in 40 CFR, Parts 144 through 149. These regulations outline the general requirements of the UIC program, the requirements for state programs, and specific standards for well construction and testing. A comparison of these regulations is summarized in Table 4.4. This table includes only those classes of injection wells that are connected with energy technologies. These classes are Class II wells (associated with oil and gas production), Class V wells (associated with geothermal energy), and Class VI wells (associated with carbon sequestration). Although Class I wells have also been proven to induce seismic events, they are excluded from this study because they have no association with energy extraction.

In practice, the well construction requirements shown above are almost always met by using standard oil and gas well construction techniques, such as setting surface casing below all underground sources of drinking water and cementing casing high above all injection horizons. This method of setting and cementing casing strings at strategic depths ensures underground sources of drinking water are protected by at least two strings of steel casing (sometimes more) and at least two barriers of cement (Figure 4.1). The ability of the tubing or casing to contain pressure is required to be continuously recorded in a Class VI well and is tested every 5 years for Class II wells.

Other governmental agencies, in addition to the EPA or a state agency, may have jurisdiction over the injection permitting process. These additional agencies include the BLM, USFS, and USGS.

TABLE 4.4 Comparison of Regulations for Wells in the EPA UIC Program

Class	II	V	VI
Use	Injection of brines and other fluids associated with oil and gas production and hydrocarbons for storage. Injected beneath the lowermost USDWs (150,851 wells).[a]	All injection wells not included in Classes I–IV. Generally used to inject nonhazardous fluids into or above USDWs and typically shallow onsite disposal systems (estimated 640,000 wells).[b] Of this number, approximately 234 wells are used for the injection of fluids in association with the recovery of geothermal energy for heating, aquaculture, and production of electric power (EPA, 1999). Permits for geothermal injection wells can be issued by the BLM in addition to state agencies or the EPA.[c]	Inject carbon dioxide (CO_2) for long-term storage, also known as geologic sequestration of CO_2 (estimated 6–10 wells by 2016).
Siting	All new Class II wells shall be sited in such a fashion that they inject into a formation which is separated from any USDW by a confining zone that is free of known open faults or fractures within the area of review (a confining zone is a formation that is capable of limiting fluid movement above an injection zone) (see 40 CFR 146.22(a)).	Class V wells have no specific siting requirements.	Owners or operators of Class VI wells must demonstrate to the satisfaction of the Director that the wells will be sited in areas with a suitable geologic system (see 40 CFR 146.83(a)). Confining zones free of transmissive faults or fracture and of sufficient areal extent and integrity to contain the injected carbon dioxide stream and displaced formation fluids and allow injection at proposed maximum pressures and volumes without initiating or propagating fractures in the confining zone(s) (see 40 CFR 146.83(a)(2)).

continued

TABLE 4.4 Continued

Class	II	V	VI
Construction	All Class II wells shall be cased and cemented to prevent movement of fluids into or between underground sources of drinking water (see 40 CFR 146.22(b)).	No specific regulations regarding well bore construction.	Surface casing must extend through the base of the lowermost USDW and be cemented to the surface through the use of a single or multiple strings of casing and cement (see 40 CFR 146.86(b)(2)). At least one long string casing ... must extend to the injection zone and must be cemented by circulating cement to the surface in one or more stages (see 40 CFR 146.86(b)(3)). All owners or operators of Class VI wells must inject fluids through tubing with a packer set a depth opposite a cemented interval at a location approved by the Director (see 40 CFR 146.86(c)(2)).
Required information	At a minimum, the following information concerning the injection formation shall be determined or calculated for new Class II wells or projects: (1) fluid pressure, (2) estimated fracture pressure, and (3) physical and chemical characteristics of the injection zone (see 40 CFR 146.22(g)). Additional information that must be considered by the Director in authorizing Class II wells includes a map showing the injection well or project area for which a permit is sought and the applicable area of review. The map may show faults if known or suspected (see 40 CFR 146.24(a)(2)).	Minimum federal UIC requirements are defined in 40 CFR 144–147. EPA Regional Offices administering the UIC program have the flexibility to establish additional or more stringent requirements based on the authorities in parts 144 through 147 (see 40 CFR 144.82(d)).	A map showing the injection well for which a permit is sought and the applicable area of review; the map should also show faults, if known or suspected; only information of public record is required to be included on this map (see 40 CFR 146.82(a)(2)). The location, orientation, and properties of known or suspected faults and fractures that may transect the confining zone(s) in the area of review and a determination that they would not interfere with containment (see 40 CFR 146.82(a)(3)(ii)). Information on the seismic history including the presence and depth of seismic sources and a determination that the seismicity would not interfere with containment (see 40 CFR 146.82(a)(3)(v)).

Operational requirements	Injection pressure at the wellhead shall not exceed a maximum which shall be calculated so as to ensure that the pressure during injection does not initiate new fractures or propagate existing fractures in the confining zone adjacent to the USDWs (see 40 CFR 146.23(a)(1)). Injection between the outermost casing protecting underground sources of drinking water and the well bore shall be prohibited.	No specific injection requirements are outlined for Class V wells except that "injection activity cannot allow the movement of fluid containing any contaminant into USDWs, if the presence of that contaminant may cause a violation of the primary drinking water standards … or may otherwise adversely affect the health of persons" (see 40 CFR 144.82(a)).	Except during stimulation, the owner or operator must ensure that injection pressure does not exceed 90 percent of the fracture pressure of the injection zone(s) so as to ensure that the injection does not initiate new fractures or propagate existing fractures in the injection zone(s) (see 40 CFR 146.88(a)).
Termination of permits	The Director may terminate a permit during its term, or deny a permit renewal application for the following cause: a determination that the permitted activity endangers human health or the environment and can only be regulated to acceptable levels by permit modification or termination (see 40 CFR 144.40(a)(3)).	The Director may terminate a permit during its term, or deny a permit renewal application for the following cause: a determination that the permitted activity endangers human health or the environment and can only be regulated to acceptable levels by permit modification or termination (see 40 CFR 144.40(a)(3)).	The Director may terminate a permit during its term, or deny a permit renewal application for the following cause: a determination that the permitted activity endangers human health or the environment and can only be regulated to acceptable levels by permit modification or termination (see 40 CFR 144.40(a)(3)).

[a] The table provided by EPA describes Class II wells as "injected below the lowermost USDW." Although this is correct in most cases, injection below the lowermost USDW is not required for Class II wells, according to UIC regulations.

[b] Most Class V wells are unsophisticated shallow disposal systems that include storm water drainage wells, cesspools, and septic system leach fields. Wells used for fluid injection in association with the recovery of geothermal resources are included in this class because they are an injection well "not included in Classes I-IV."

[c] A review of geothermal injection wells performed by the EPA in 1999 notes, "The permits [for Class V wells] are issued by state agencies, US Bureau of Land Management (BLM), and/or the USEPA Regional Office, depending on the state and whether the well is located on state, federal, or private land. In general, the permits are similar to those issued for Class II injection wells" (EPA, 1999).

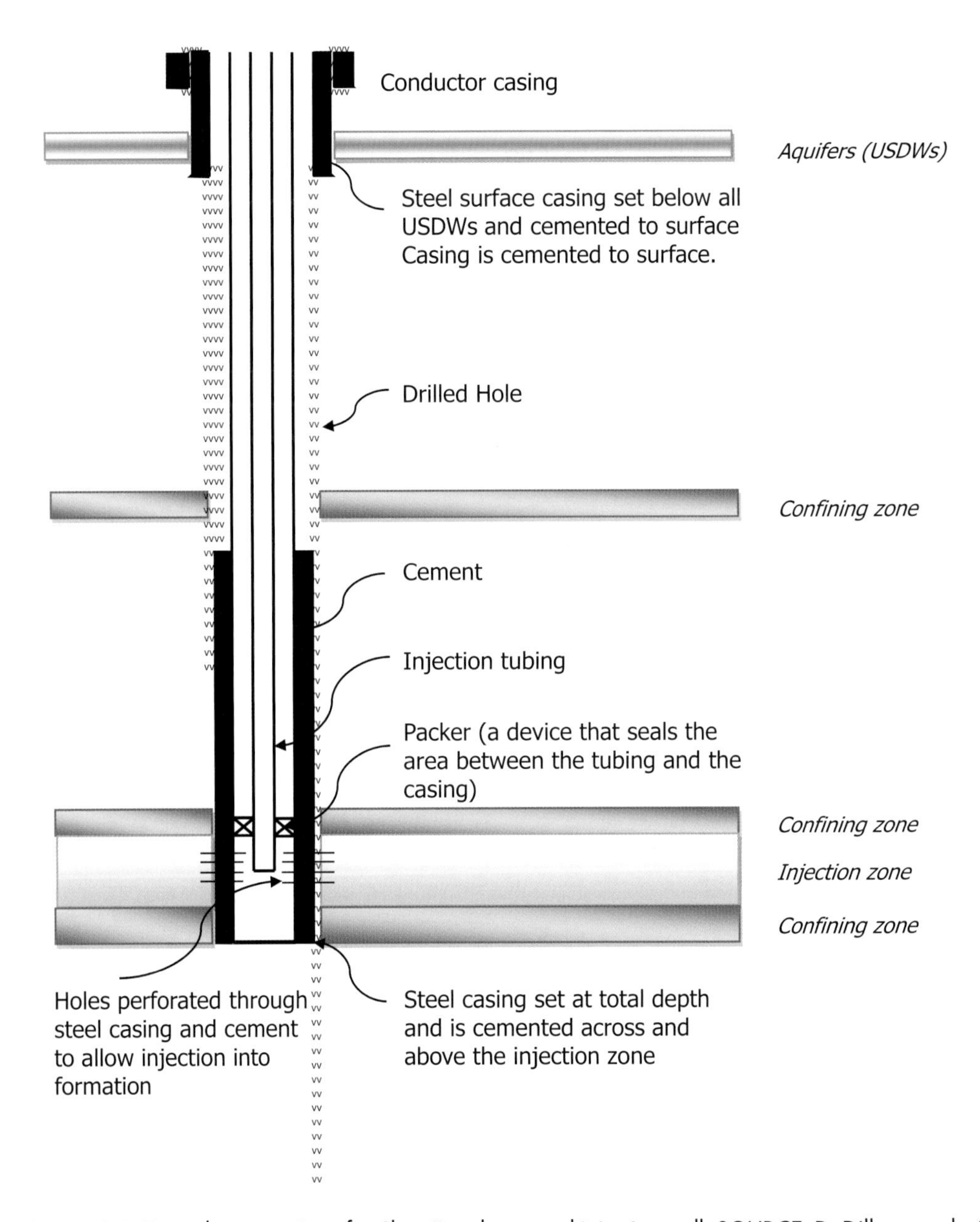

FIGURE 4.1 Typical construction of a Class II underground injection well. SOURCE: D. Dillon, used with permission.

Bureau of Land Management

The BLM has jurisdiction over onshore leasing, exploration, development, and production of oil and gas on federal lands in the United States.[7] Certain contractual property rights and responsibilities governing resource development are created when BLM issues a lease to extract oil and gas resources or geothermal energy from federal lands (NRC, 2010). The BLM regulatory framework governing oil and gas extraction operations for federal and tribal lands is contained in 43 CFR Part 3160 (Onshore Oil and Gas Operations).[8] In the process of underground injection, the BLM normally has the role of a surface owner with jurisdiction over surface facilities and surface impacts. (For example, the BLM is required to take National Environmental Policy Act [NEPA] provisions in its management of surface resources.) Permitting and construction considerations for these wells are reviewed and approved by the EPA or the appropriate state regulatory agency. The permitting and oversight of geothermal wells, however, can be an exception. The "Geothermal Steam Act" (43 CFR Parts 3200, 3210, 3220, 3240, 3250, and 3260) gives the BLM authority to regulate geothermal resources on federal lands administered by the Department of the Interior and the Department of Agriculture, where geothermal resources were reserved to the United States. In these cases the BLM permits, approves, and regulates the development of geothermal resources (Box 4.1).

U.S. Forest Service

The USFS is primarily responsible for managing surface resources on national forest lands. The USFS cooperates with the Department of the Interior in administering exploration and development of leasable minerals, including the review of permit and lease applications and making recommendations to protect surface resources (USFS, 1994). As is the case with BLM, the USFS takes the role of surface owner in injection activities and exercises jurisdiction over surface facilities and surface impacts that are associated with injection operations. The USFS is also required to take into account NEPA provisions in its management of surface resources. The actual permitting and oversight of injection activities is exercised by the EPA, local state agencies, or the BLM.

[7] BLM is primarily responsible for the regulation and development of federal oil and gas mineral resources under the following acts: the Mining Leasing Act of 1920 (41 Stat. 437; see BLM, 2007); the Federal Land Policy and Management Act of 1976 (43 USC 1701-1782; see BLM, 2001); the Federal Onshore Oil and Gas Leasing Reform Act of 1987 (101 Stat. 1330-256, an amendment to the Mineral Leasing Act of 1920); the National Forest Management Act (16 USC 1600-1604); and the National Materials and Minerals Policy, Research, and Development Act of 1980 (P.L. 96-479; 30 USC 1601-1605). Many of these acts are summarized in NRC (1989).

[8] The BLM and USFS jointly prepared a manual, *The Gold Book*, which summarized surface operating standards and guidelines for oil and gas exploration and development (BLM and USFS, 2007).

BOX 4.1
BLM Regulation of Class V Geothermal Injection Wells: Seismicity Concerns

The BLM, through an informal agreement with the EPA, regulates the Class V geothermal injection wells in California. Under this arrangement the BLM has recently issued its "Conditions of Approval" for a proposed enhanced geothermal systems project that stipulated the specific procedures to be followed in the event that induced seismicity is observed to be caused by the proposed stimulation (hydraulic fracturing) operation.[a] As issued by the BLM, the specific procedures include the use of a "traffic light" system that allows hydraulic fracturing to proceed as planned (green light) if it does not result in an intensity of ground motion in excess of Mercalli IV "light" shaking (an acceleration of less than 3.9%g), as recorded by an instrument located at the site of public concern. However, if ground motion accelerations in the range of 3.9%g to 9.2%g are repeatedly recorded, equivalent to Mercalli V "moderate" shaking, then the hydraulic fracturing operation is required to be scaled back (yellow light) to reduce the potential for a further occurrence of such events. Finally, if the operation results in producing a recorded acceleration of greater than 9.2%g, resulting in "strong" Mercalli VI or greater shaking, then the active hydraulic fracturing operation is to immediately cease (red light).

[a] R.M. Estabrook, BLM, Conditions of Approval for GSN-340-09-06, Work Authorized: Hydroshear, The Geysers, January 31, 2012.

U.S. Geological Survey

The USGS provides scientific information to describe and understand the Earth; minimize loss of life and property from natural disasters; manage water, biological, energy, and mineral resources; and enhance and protect quality of life in the United States.[9] It is the only federal agency with responsibility for recording and reporting earthquake activity worldwide, and it is often asked to aid state agencies in the investigation of possible induced seismicity. Its Earthquake Hazards Program serves as the USGS component of the multiagency National Earthquake Hazards Reduction Program, which develops, disseminates, and promotes knowledge, tools, and practices for earthquake risk reduction that improve national earthquake resilience. The Earthquake Hazards Program also houses the National Earthquake Information Center (NEIC), which aims to determine the location and size of all destructive earthquakes worldwide and to disseminate this information to concerned agencies, scientists, and the general public.

The USGS is continuing to enhance its earthquake monitoring and reporting capabilities through the Advanced National Seismic System (ANSS). Since 2008 the USGS has

[9] See www.usgs.gov/aboutusgs/.

installed approximately 300 new earthquake-monitoring instruments in the highest-risk areas. Full implementation of ANSS will result in 6,000 new instruments on the ground and in structures in at-risk urban areas (Box 4.2).

Seismic events that are thought to be induced are flagged in the USGS earthquake database. However, many or most events that USGS scientists suspect may be induced are not labeled as such, due to lack of confirmation or evidence that those events were in fact induced by human activity.[10] This is often true with events in regions that have experienced natural earthquakes before any mining or extraction operations were established. The earthquake location accuracy provided by the NEIC depends primarily on the number and location of seismic stations recording the event. During the 2008-2009 Dallas–Fort Worth earthquake swarm, the accuracy of the initial NEIC locations was on the order of 10 km (6 miles), which made the events difficult to assign to a particular injection well (Frohlich et al., 2011). In areas of low historical seismicity, the NEIC network coverage tends to be sparser than in more seismically active areas, making the detection of small events (< **M** 3) and accurate hypocenter locations difficult (Box 4.3).

STATE EFFORTS

Although the concerns surrounding induced seismicity are relatively new, at least two states have now adopted, or are in the process of adopting, regulations or approval procedures to address the issue. Colorado and Arkansas are currently reviewing underground injection permits for possible problems with induced seismicity in the Raton Basin, Colorado, and Guy-Greenbrier area, Arkansas (Box 4.4). Recent seismic activity in the Raton Basin near a large coalbed methane field with active injection has prompted the Colorado Oil and Gas Conservation Commission (COGCC) to initiate a policy requiring the Colorado Geologic Survey to review all Class II injection permits for geologic features that could result in seismicity due to injection. According to a statement released by the COGCC, "if historical seismicity has been identified in the vicinity of a proposed Class II UIC well, COGCC requires an operator to define the seismicity potential and the proximity to faults through geologic and geophysical data prior to any permit approval" (COGCC, 2011). Due to apparent instances of induced seismicity in Arkansas, the Arkansas Oil and Gas Commission (AOGC) proposed regulations to establish a "Moratorium Zone" covering over 1,000 square miles where no permit for a Class II well will be granted without a hearing by the Commission (AOGC, 2012). The proposed regulations also require no Class II permit will be issued within 5 miles of a "Moratorium Zone Deep Fault" without a hearing by the Commission.

[10] Bruce W. Presgrave, USGS, personal communication, March 3, 2011.

BOX 4.2
Temporary Seismic Array Acquisition and Processing Cost Estimates

In the event of a felt induced seismic event, a temporary seismic network may be installed to augment the regional network or to record the events within the temporary network. This involves installing sensitive seismic instruments around the area of interest to record small earthquakes that are typically difficult to detect on more than a few instruments within a standard regional array. By augmenting the regional seismic stations with a dense temporary seismic network, seismologists can carry out detailed analyses on the earthquake waveforms and improve the earthquake location accuracy in the subsurface. Additionally, if the data and station coverage around an induced seismic event is appropriate, a better understanding of the earthquake's size and failure mechanism can be determined. The cost of a temporary seismic array including the array deployment, operation, and data analyses will depend on the number of stations, the location of the study area, the length of the study period, and the overall goals of the seismic monitoring project.

A variety of instruments is commercially available for recording small earthquakes. Broadband instruments specialize in recording a broad spectrum of waveforms from 120 to 175 Hz. Short-period instruments are equipped to record only high frequencies, in general > 1 Hz. A complete broadband station with recorder, geophone, and assorted auxiliary equipment costs around $25,000, and a short-period recorder is slightly less at approximately $20,000 (2011 cost estimates). Eight to 10 instruments are typically deployed for a small temporary seismic array, but as many as 20 instruments are deployed for more detailed earthquake surveys. The network sensitivity is often measured by how small an event can be recorded and located, and the array design will depend on sensitivity required for the study (for example, to record and locate an event down to **M** 0). Hence, for a temporary seismic array, instrumentation costs alone run from $120,000 to $370,000. The seismic instruments can be reused after the study is completed; some minor costs are associated with instrument maintenance and storage.

The expenditure associated with installing and running the temporary seismic array will depend on the location of the array and the length of the deployment. Estimated costs for a 150-day deployment are approximately $100,000, which includes the mobilization, demobilization, equipment setup, tie in with existing seismic network, and charges for data telemetry. The seismic instrumentation is very sensitive to ground motion, and geophones cannot be installed in areas with high background noise, such as freeways, busy urban areas, factories, etc., as they will be saturated with noise and unable to record seismic signal from small earthquakes. In noisy areas the seismic instruments may have to be placed in shallow boreholes (typically 200 to 400 feet deep), which will add additional cost to the array installation, which is not included in the price listed above.

Detailed analysis of the seismic data by qualified seismologists is required to determine earthquake hypocenters and magnitudes, estimate location errors, and determine the type of failure (focal mechanism or moment tensor inversion). The cost for the work will depend on the detail required; cost estimates for professional analysis for a 6-month seismic deployment is in the $200,000 to $300,000 range for a university-based project. Commercial companies—national laboratories, for example—are available to provide these types of services and prices will vary depending on the project scope. Thus the total cost, including purchasing seismic instruments and installing and operating the array for a 150-day deployment with 8 to 12 instruments is estimated at $400,000 to $800,000.

Less costly recording instruments are being developed that could significantly drive down the cost of an instrument to less than $1,000 to $3,000 per site (Hutchings et al., 2011); however, the type of instrumentation used will depend on the goal of the study. Overall instrumentation is a minor cost compared to the overall deployment and interpretation of the seismic data.

BOX 4.3
The National Earthquake Information Center

The NEIC,[a] headquartered in Golden, Colorado, is responsible for quickly determining the location and size of destructive earthquakes worldwide and disseminating in near real time the information to concerned national and international agencies, scientists, and the general public. NEIC produces a comprehensive catalog of earthquake source parameters and macroseismic effects for all **M** 4.5+ earthquakes worldwide and **M** 2.5+ earthquakes in the United States in coordination with USGS-supported regional seismic networks (see Figure for a map of magnitude sensitivity of the seismic network within the United States). The NEIC acquisition and processing system is designed for recording and analyzing seismic earthquakes on all scale lengths from near-real-time monitoring of aftershock sequences using dense local arrays to modeling of all damaging earthquakes worldwide.

For example, NEIC in 2011 simultaneously and seamlessly reported on the 2011 **M** 9.0 Japanese earthquake and its aftershocks and multiple earthquake sequences in the United States that included Guy, Arkansas; Mineral, Virginia; Prague, Oklahoma; and Trinidad, Colorado. In later cases, the existing seismic monitoring system was augmented by dense local seismic stations that enabled automatic detection and locations to magnitudes less than about 1.5.

In addition, NEIC and the Earthquake Hazards Program maintains a group of 32 portable seismic recording systems, designed for both strong (large earthquakes) and weak motion (events less than **M** 3), in order to respond to notable seismic sequences throughout the United States. This equipment is often loaned to

Box continues

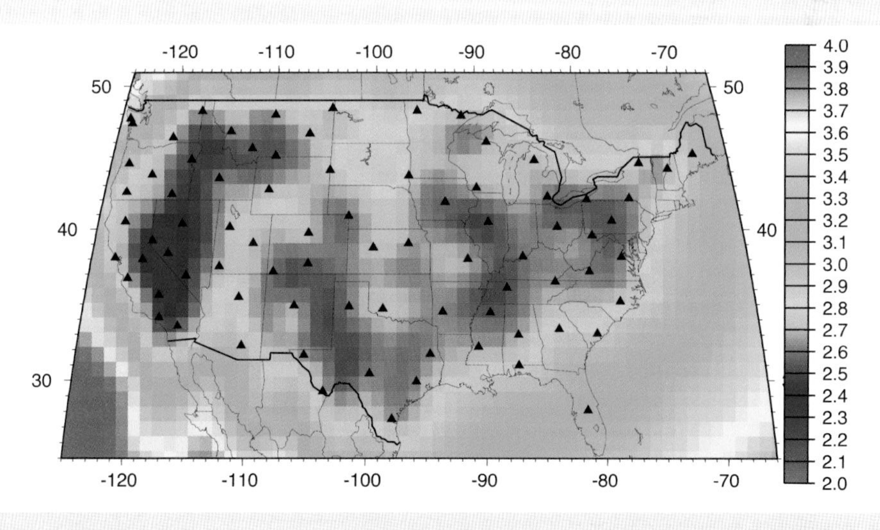

Figure Map of the minimum detectable earthquake magnitude within the lower 48 states using the ANSS array operated by the USGS/NEIC. Shading indicates the minimum-sized earthquake that can be detected and located by the NEIC, as indicated by the color bar on the right. Triangles mark seismic station locations. SOURCE: USGS/NEIC.

BOX 4.3 Continued

cooperating state geological surveys and regional seismic networks to address specific local seismic monitoring issues. NEIC's acquisition and processing system allows them to automatically integrate near-real-time and non-real-time waveform and source parameter data from regional seismic networks and portable seismic stations to develop complete seismic catalogs of earthquake sequences. As an example, NEIC is presently integrating its existing seismic bulletin for the 2011 **M** 5.8 Virginia earthquake with other non-real-time data from more than 40 stations deployed by multiple universities and/or state and federal agencies.

[a] See earthquake.usgs.gov/regional/neic/.

BOX 4.4
The 2010-2011 Guy Greenbrier, Arkansas, Earthquake Swarm and Arkansas Class II Injection Well Moratorium Area

A group of Class II wastewater disposal wells started operation April 2009 in central Arkansas, near the towns of Guy and Greenbrier, Arkansas. The wells were used to dispose of wastewater associated with gas development from the Fayetteville shale. A swarm of earthquakes (**M** ≤ 4.7) started in September 2010 between the towns of Guy and Greenbrier (Figure 1). The close spatial and temporal correlation between the seismicity and the wastewater injection wells suggests a link between injection and seismicity. All but 2 percent of the earthquake activity occurred in the vicinity of about a 6-km (3.7-mile) radius of three specific injection wells (labeled wells 1, 2, and 5 in Figure 1) (Horton, 2012). One injection well, number 5, appears to intersect a known fault, the Enders, which may allow fluid to travel down into deeper crustal structures (Horton, 2012).

Central Arkansas commonly experiences diffuse swarm seismicity, which is thought to be associated with the New Madrid Seismic Zone (NMSZ), the largest seismic zone east of the Rocky Mountains. The NMSZ is located on the northeastern part of Arkansas, southeast Missouri, and northwest Tennessee (Figure 1). The Guy-Greenbrier area has a history of seismic activity, a series of earthquakes referred to as the Enola swarms, which occurred in the 1980s and in 2001, east-southeast of Greenbrier (Figure 2). The Enola swarms were not well located due to poor instrumentation; however, the activity tended to form elongated east-west trends from 3 to 7 km (9850–23,000 feet) in depth (Chui et al., 1984; Rabak et al., 2010).

The AOGC approved a moratorium for any new or additional Class II disposal on January 26, 2011. The injection moratorium area is approximately 5 miles surrounding the Guy-Greenbrier and Enola seismically active area and covers an area of over 1,150 square miles (AOGC, 2011).[a] Operators with existing Class II wells are required to report daily injection pressures and volumes to the AOGC director. The moratorium

Box continues

BOX 4.4 Continued

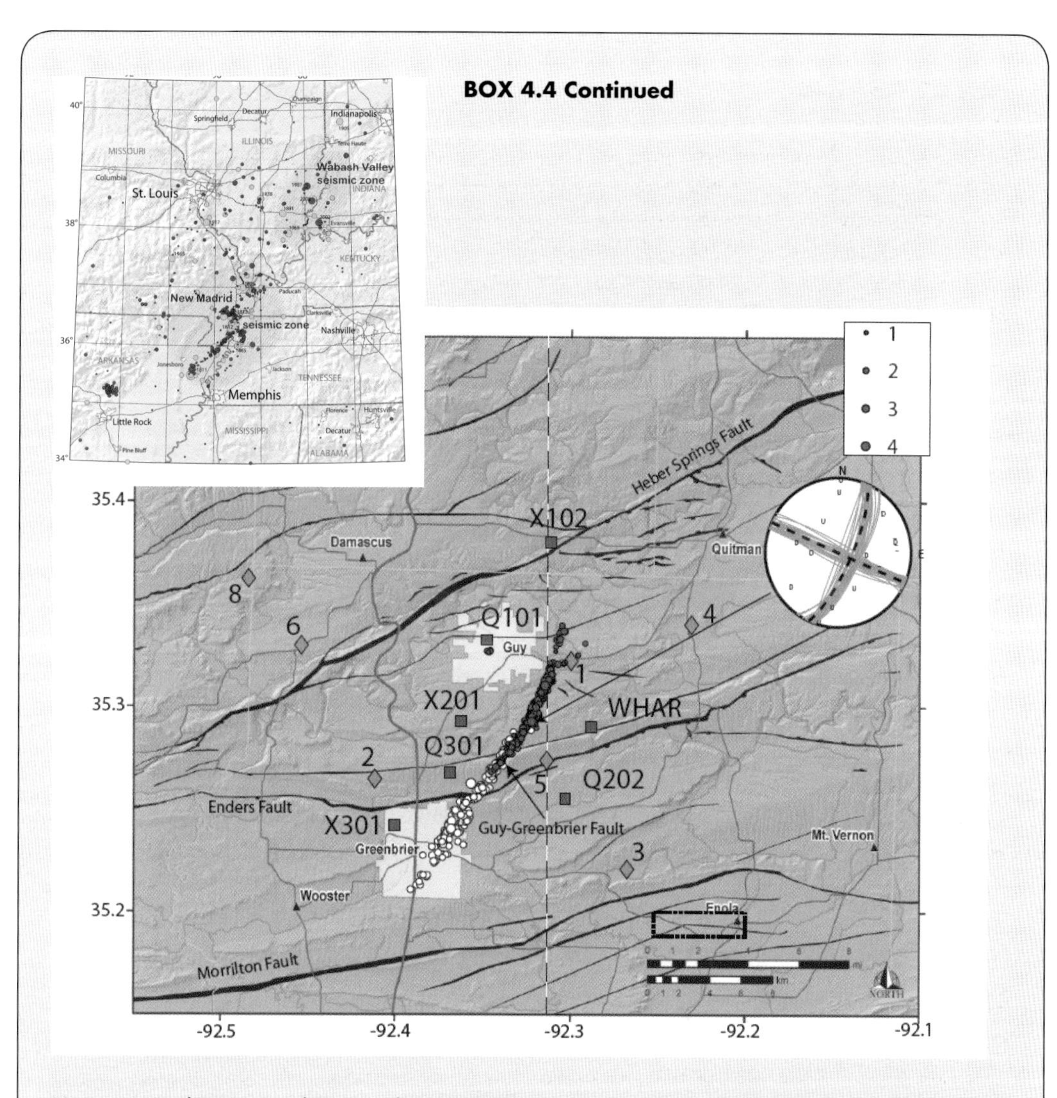

Figure 1 Study area in Arkansas with Guy-Greenbrier seismic activity. Seismic stations installed by the Arkansas Geological Survey and the Center for Earthquake Research at the University of Memphis are marked by black squares; injection wells are marked by red dots; seismic events between October 1, 2010, and February 15, 2011, are marked by dark gray dots; and seismic events between February 16, 2011, and March 8, 2011, are marked by white dots. Named faults penetrate to the Precambrian basement (faults from AGS and AOGC). Right inset: First-motion focal mechanism for M 4.0 event on October 11, 2010, is consistent with right-lateral strike slip on a northeast-oriented fault. Left inset shows the location of the New Madrid Fault zone in northeastern Arkansas with historical earthquakes. SOURCE: Modified from Horton (2012); see also earthquake.usgs.gov/earthquakes/eqarchives/poster/2011/20110228.php.

Box continues

BOX 4.4 Continued

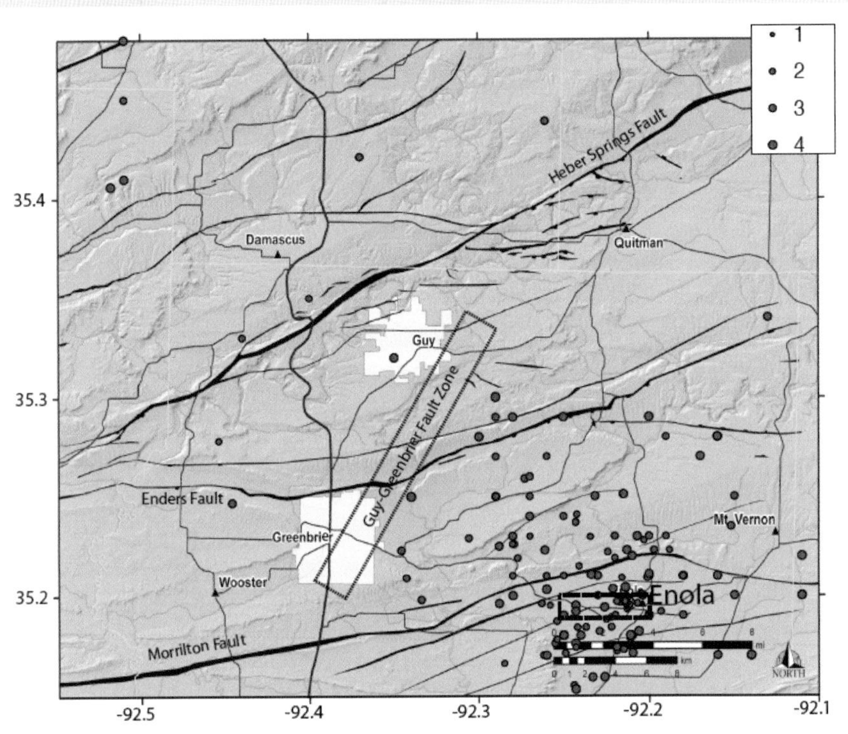

Figure 2 Map of the historical seismicity in the Guy-Greenbrier area 1976 to 2009, including the Enola swarm. SOURCE: Horton (2012).

was placed to allow time to investigate "a potential correlation between the seismic activity and disposal well operations in the Guy-Greenbrier, Arkansas area" (AOGC, 2011). In the surrounding Fayetteville shale development area outside the Permanent Moratorium Area, the AOGC director may propose additional requirements for any new disposal wells (AOGC, 2011).

[a] See AOGC (2011) for a detailed map of the AOGC's proposed Permanent Moratorium Area for disposal wells.

EXISTING REGULATORY FRAMEWORK FOR FLUID WITHDRAWAL

While the injection of fluid underground is regulated by the EPA, the BLM, and state agencies, the extraction of fluids is normally not regulated or is minimally regulated. The number of events of induced seismicity caused by the withdrawal of fluid is approximately equal to the number of events caused by the underground injection of fluids for both disposal and secondary recovery (see Box 1.1 in Chapter 1), but fluid withdrawal is usually not curtailed due to induced seismicity. This is because the pumping of fluids from underground reservoirs can be divided among many different oil companies, and states only require permits to drill oil and gas wells, not to produce fluid from them. One method of controlling the withdrawal of fluids from an underground reservoir is through "unitization." Unitization is an order granted by the state oil and gas regulators that designates one oil and gas company to be the "unit operator" of the unitized oil and gas field, and profits and expenses from oil and gas operations are divided among operators as dictated by the unitization agreement (Box 4.5). This

BOX 4.5
Unitization

In 1892 Edward Doheny and Charlie Canfield discovered the Los Angeles City oil field. By 1895 the field had produced 729,000 barrels of oil, nearly 60 percent of California's production. The discovery was in a townlot area composed of small residential lots. Each townlot lot owner had both surface and mineral rights. California has the "Law of Capture," which means that a "liquid mineral" can move from one property to another. Because of this each owner had to drill a well or have their oil taken by their neighbor. This resulted in runaway drilling and very inefficient and expensive oil operations. Some producers would overproduce their wells and harm the productivity of their neighbors, resulting in inefficient and expensive development.

Early in the 1900s the State of California formed the Division of Oil and Gas (now called DOGGR). In the period from 1923 to 1926, Union, Shell and Associated Oil Company under a cooperative agreement developed the Dominguez oil field. This unit proved to be an efficient way to manage the field with almost no wastage. The Subsidence Control Act of 1958 encouraged voluntary pooling and unitization and provided for compulsory unitization if needed. The individual operators of an oil field would be combined into a Unit and the oil field would be operated by one party called the Unit Operator. The other participants are called Working Interest Owners. Unit documents would define the unit and the participant's share of the total, called the Equity Determination.

Unitization has proven to be an effective way to share the wealth and operations of oil fields fairly while protecting the environment and guaranteeing energy conservation. It is also easier to regulate because all parties share the profits and losses but one party, the Unit Operator, is in charge. DOGGR has used unitization to force efficient waterfloods and prevent environmental problems.

SOURCE: Rintoul (1990).

order is normally requested prior to initiating secondary recovery operations and is granted with the consent of the majority of the affected oil and gas operators. Because a unitization order is granted by a state's oil and gas governing body, it can also include requirements to limit fluid withdrawal for a variety of reasons. These might include conservation of the oil and gas resource, limited injection of fluid, or induced seismic events. Although many oil and gas fields have been unitized in the United States, we know of no instance where produced fluid volumes have been curtailed to limit induced seismicity.

CONCLUDING REMARKS

Although the SDWA provides a regulatory framework for the underground injection of fluids, the act does not explicitly address the issue of induced seismicity or how induced seismic events should be investigated and regulated. Currently, many different agencies have oversight of the UIC program, such as the EPA, various state agencies, the BLM, and the USFS. To date, these various agencies have dealt with induced seismic events with different and localized actions, using input from additional government agencies such as the USGS and various state geologic surveys, as well as university researchers. These efforts to respond to incidence of perceived induced seismicity have been successful but are of an ad hoc nature and can vary widely depending on the different agencies involved.

REFERENCES

AOGC (Arkansas Oil and Gas Commission). 2011. Proposed Permanent Moratorium for Disposal Wells in Certain Areas. Public Announcement—Docket No. 180A-2011-07. January 26. Available at stream.loe.org/images/110624/Proposed_Permanent_Moratorium.pdf (accessed October 2012).

AOGC. 2012. Final Rule H-1—Class II Disposal and Class II Commercial Disposal Well Permit Application Procedures. February 17. Available at aogc.state.ar.us/PDF/H-1%20FINAL%202-17-2012.pdf (accessed October 2012).

BLM (Bureau of Land Management). 2001. The Federal Land Policy and Management Act of 1976, as Amended. Washington, DC: U.S. Department of the Interior. Available at www.blm.gov/or/regulations/files/FLPMA.pdf (accessed December 2012).

BLM. 2007. Mineral Leasing Act of 1920 as Amended. Retranscribed on August 9, 2007. Washington, DC: U.S. Department of the Interior. Available at www.blm.gov/or/regulations/files/mla_1920_amendments1.pdf (accessed December 2012).

BLM and USFS. 2007. *Surface Operating Standards and Guidelines for Oil and Gas Exploration and Development: The Gold Book*, 4th ed. Washington, DC: U.S. Department of the Interior. Available at www.blm.gov/pgdata/etc/medialib/blm/wo/MINERALS__REALTY__AND_RESOURCE_PROTECTION_/energy/oil_and_gas.Par.18714.File.dat/OILgas.pdf (accessed January 2012).

Chui, J.M., A.C. Johnston, A.G. Metzger, L. Haar, and J. Fletcher. 1984. Analysis of analog and digital records of the 1982 Arkansas earthquake swarm. *Bulletin of the Seismological Society of America* 74:1721-1742.

COGCC (Colorado Oil and Gas Conservation Commission). 2011. COGCC Underground Injection Control and Seismicity in Colorado. January 19. Denver, CO: Department of Natural Resources. Available at cogcc.state.co.us/Library/InducedSeismicityReview.pdf (accessed December 2012).

EPA (U.S. Environmental Protection Agency). 1999. The Class V Underground Injection Control Study. Volume 17: Electric Power Geothermal Injection Wells. EPA/816-R-99-014q. Washington, DC. Available at www.epa.gov/ogwdw000/uic/classv/pdfs/volume17.pdf (accessed November 2012).

EPA. 2010. 2010 UIC Well Inventory. EPA Underground Injection Control Program. Available at water.epa.gov/type/groundwater/uic/upload/UIC-Well-Inventory_2010-2.pdf (accessed February 2012).

Frohlich, C.F., C. Hayward, B. Stump, and E. Potter. 2011. The Dallas–Fort Worth earthquake sequence: October 2008 through May 2009. *Bulletin of the Seismological Society of America* 101(1):327-340.

Horton, S. 2012. Disposal of hydrofracking-waste fluid by injection into subsurface aquifers triggers earthquake swarm in central Arkansas with potential for damaging earthquake. *Seismological Research Letters* 83(2):250-260.

Hutchings, L.J., B. Jarpe, K.L. Boyle, B.P. Bonner, G. Viegas, H. Philson, P. Statz-Boyer, and E. Majer. 2011. Near-real time, high resolution reservoir monitoring and modeling with micro-earthquake data. Presented at the American Geophysical Union Fall 2011 Meeting, San Francisco, CA, December 5-9.

NRC (National Research Council). 1989. *Land Use Planning and Oil and Gas Leasing on Onshore Federal Lands.* Washington, DC: National Academy Press.

NRC. 2010. *Management and Effects of Coalbed Methane Produced Water in the United States.* Washington, DC: The National Academies Press.

Rabak, I., C. Langston, P. Bodin, S. Horton, M. Withers, and C. Powell. 2010. The Enola, Arkansas, intraplate swarm of 2001. *Seismological Research Letters* 81:549-559.

Rintoul, W. 1990. *Drilling Through Time, 75 Years with California's Division of Oil and Gas.* Division of Oil and Gas Publication TR40. Sacramento, CA: California Department of Conservation.

USFS (U.S. Forest Service). 1994. Forest Service Manual, Title 2800—Minerals and Geology. Amendment No. 2800-94-2. Available at www.fs.fed.us/im/directives/fsm/2800/2820.txt (accessed December 2012).

CHAPTER FIVE

Paths Forward to Understanding and Managing Induced Seismicity in Energy Technology Development

Induced seismicity has associated hazards and risks that can, in concept, be quantified. Understanding what is meant by "hazard" and "risk" related to induced seismicity is critical to any discussion of the options that can be employed to mitigate the possibility of felt induced seismicity and potential impacts from development of energy technologies. To promote a better understanding of hazards and risks, we first define these terms precisely and identify the factors that influence them. The remainder of the chapter discusses hazards and risks associated with induced seismicity and steps that can be taken to quantify hazard and risk associated with induced seismicity. The committee envisions future approaches toward mitigation of any hazards associated with induced seismicity involving "best practices" protocols as a cooperative endeavor between industry, government, and the public (Chapter 6).

HAZARDS AND RISKS ASSOCIATED WITH INDUCED SEISMICITY

Definitions

The *hazard of induced seismicity* is the description and possible quantification of what physical effects could be generated by human activities associated with subsurface energy production or carbon capture and storage (CCS). For this discussion, physical effects include microseisms, earthquakes, and the associated ground shaking, both underground and at the Earth's surface. In concept it is possible to calculate probabilities of the occurrence of microseisms and earthquakes and, given one of these events, to predict the possible ground motions. However, making such calculations requires assembling statistical data that are not readily available, such as the total number of wells of different depths, the geologic environments (including faults and plate motions), production characteristics from the well(s), and the subsets of those wells that generate microseisms and earthquakes of various magnitudes.

The *risk of induced seismicity* is the description and possible quantification of how induced earthquakes might cause losses (damage to structures, and effects on humans including injuries and deaths). The losses generally occur on the Earth's surface, although underground losses, for example damage to nearby petroleum wells, could also be analyzed. The concept of *risk* involves predicting the effect of induced ground motions, and perhaps fault slip, on structures and humans. If structures can incur moderate or heavy damage, *risk* involves predicting the effect of that damage (e.g., structural collapse) on humans in the vicinity.

Note that *risk* involves loss caused by structural damage, including effects on humans. If no structures or other constructed facilities are present, for example because the causative earthquakes occur in an uninhabited area, there is no risk. (Exceptions always exist to these general statements. One case would be an earthquake causing a rock slide that injures hikers in a national park, with no structure involved, but such cases would be rare exceptions.) The concept of *risk* could also be extended to include frequently occurring ground shaking that is a *nuisance* to humans (in the general, rather than legal, sense).

Factors Affecting Hazard

A set of questions can be addressed to understand and possibly quantify the hazard and risk associated with induced seismicity associated with energy technologies (Figure 5.1). Descriptions of each question are as follows:

1. Does an energy technology at a particular location generate apparent seismic events (meaning those that are felt at the surface)? The large majority of activities associated with hydrocarbon production do not cause any apparent seismic events. If no seismic events are recorded, this may be because the seismic events are too small (e.g., **M** < 0.0) to be recorded by regional seismic instruments, but the effect is the same: there is no apparent seismic activity.
2. Does an energy technology at a particular location generate just microseisms, or microseisms and earthquakes? This question involves the size of seismic events that are associated with the energy technology. Microseisms (by definition, seismic events with **M** < 2.0) generally do not produce ground motions strong enough to have an effect on structures, but they can in cases of close proximity be felt by humans at the surface. For example, two shallow (~2 km deep [1.2 mile deep]) seismic events of **M** 1.5 and **M** 2.3 in Blackpool, England, were reported by a number of people to have been felt in April and May 2011 (BGS, 2011).
3. Can earthquake shaking be felt at the surface? Not all earthquakes are felt at the surface. Earthquake ground motions at the surface depend on the size (magnitude) of the event and its depth, among other factors. The deeper the earthquake, the larger it must be to cause ground motions at the surface that can be felt by humans. Very shallow seismic activity (e.g., 2 km) has a higher hazard of causing felt ground motions than deeper activity.

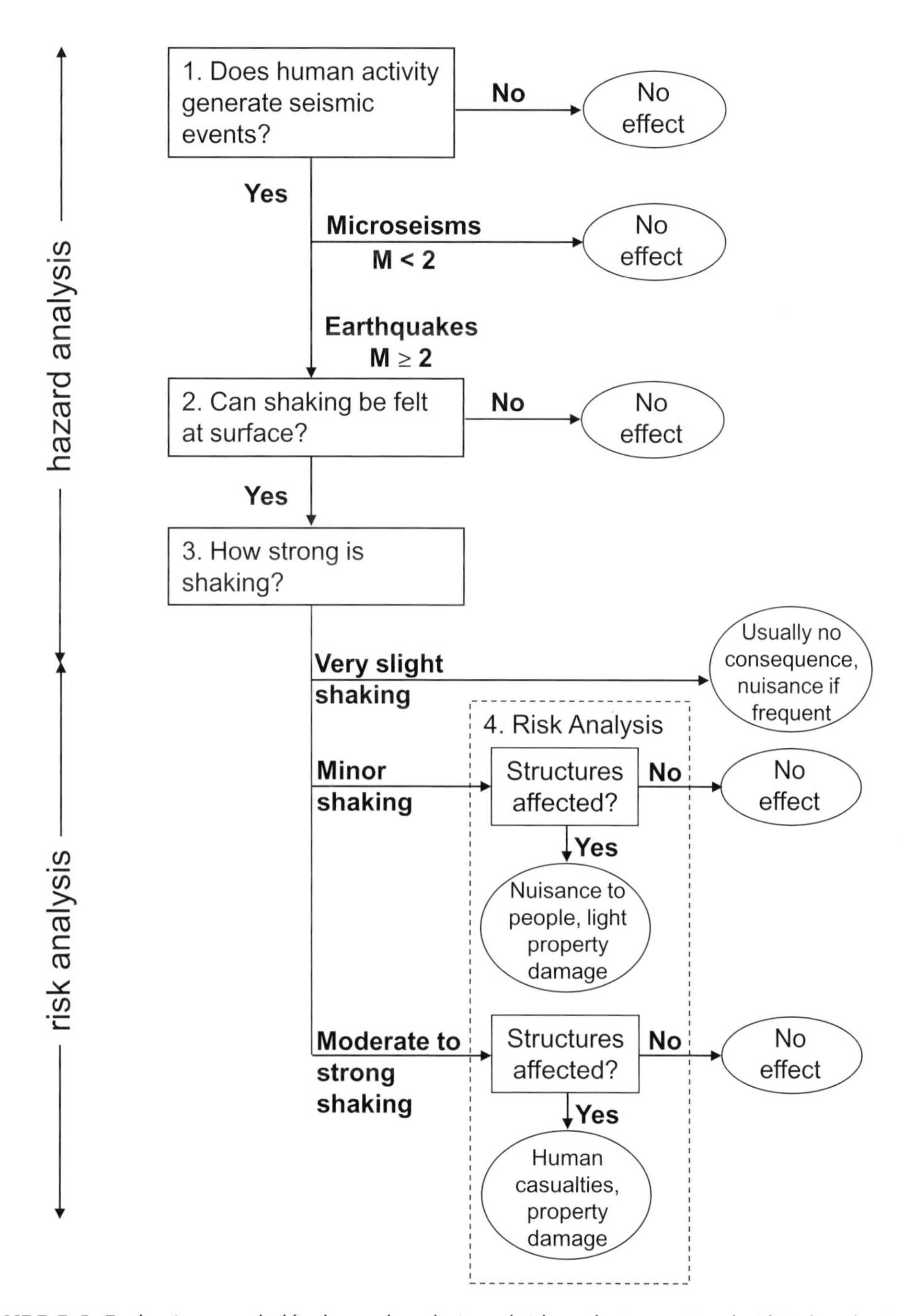

FIGURE 5.1 Evaluations needed for hazard analysis and risk analysis associated with induced seismicity for one well.

4. What are the shaking effects? If ground motions are strong enough to be felt, they can be represented by three categories, depending on the maximum strength of shaking. Ground motions fall into the following categories:

   a. Very slight shaking. These are felt ground motions, typically with peak accelerations less than 4%g, and do not cause damage to structures. Isolated cracks in plaster walls may be observed or items in houses may be knocked over, but these motions cause no damage of consequence. Frequent occurrence of these motions may be a nuisance to people.
   b. Minor shaking. These are ground motions that frighten people and/or wake them from sleep, typically with peak accelerations between about 4%g and 18%g. If structures are present, these ground motions may cause light property damage (cracks in concrete, broken windows, or cosmetic damage) but do not cause buildings to collapse.
   c. Moderate-strong shaking. These are moderate, strong, or severe ground motions with the potential of causing moderate or heavy damage, typically with peak accelerations greater than about 18%g. If structures are present, moderate to heavy damage may occur, including partial or complete collapse of structures or structural elements (foundations, walls, roofs). These effects on structures may cause human casualties (injuries and deaths, in severe cases).

Ground motions from induced seismicity generated at shallow depths can be more troublesome compared to the ground motions from deeper events (Figures 5.2 and 5.3). In a cross section of the Earth where a deep tectonic (natural) earthquake occurs at a depth of 10 km (Figure 5.2), the semicircles illustrate the distance within which minor shaking (or greater) occurs if the earthquake **M** is 3, 4, or 5. Because of the depth of the earthquake, minor (or greater) shaking usually does not reach the Earth's surface for **M** 3 or 4. For **M** 5, minor (or greater) shaking may occur at the Earth's surface within about 15 km (9 miles) of the epicenter (Figure 5.2).

Figure 5.3 shows a similar cross section of the Earth where a shallow earthquake occurs at the bottom of a 2-km-deep (1.2-mile-deep) well. Because of this shallow depth, a **M** 4 earthquake can cause minor (or greater) shaking within about 8 km of the well, and a **M** 3 earthquake may cause minor (or greater) shaking very close to the well.

Factors Affecting Risk

Risk from induced seismicity only occurs if structures are present that may be damaged. Risk exists to those structures only if the shaking is minor, moderate, or larger. Factors that should be considered for risk include location of faults, location of infrastructure that can be

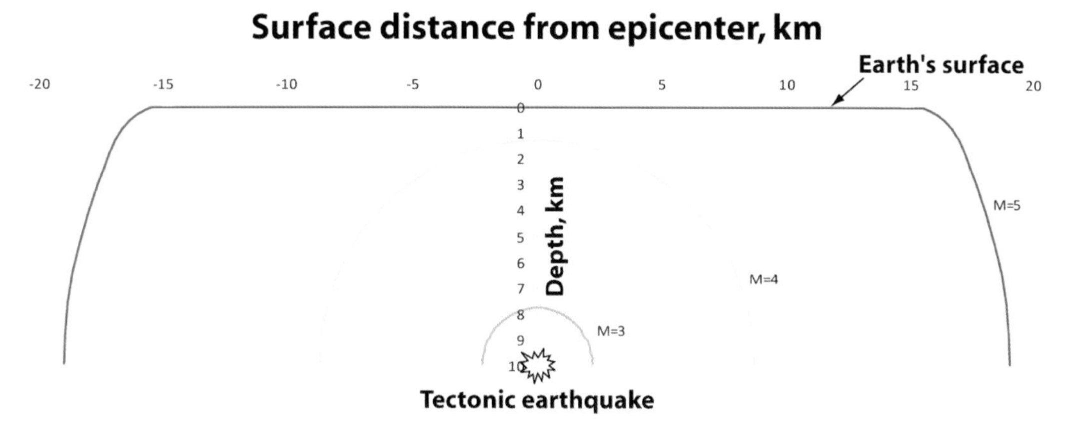

FIGURE 5.2 Cross section of the Earth illustrating the maximum distance that minor (or greater) shaking will occur, for tectonic earthquakes originating at 10 km (6 miles) depth, with **M** 3 (green line), 4 (yellow line), and 5 (red line). In this example, only **M** 5 earthquakes will generate shaking that is felt at the surface.

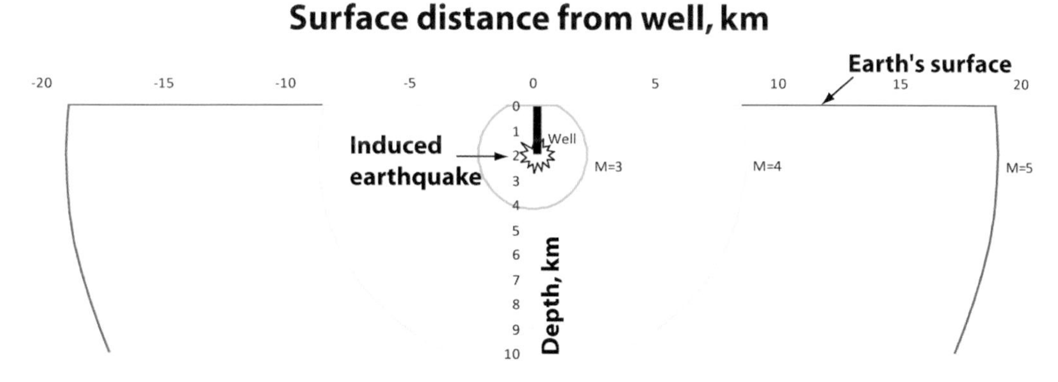

FIGURE 5.3 Cross section of the Earth illustrating maximum distance that minor (or greater) shaking will occur, for both natural and induced earthquakes originating at 2 km (1.2 miles) depth, with **M** 3 (green line), 4 (yellow line), and 5 (red line). The diagram depicts an induced earthquake at the bottom of a well. Because of the shallow depth, each of these earthquake magnitudes would generate shaking at the surface that could be felt. Because of the larger energy released, a **M** 5 earthquake would be felt over a much greater area of the surface (up to 20 km [12 miles]) from the well, whereas a **M** 3 earthquake would only be felt about 1 km (0.6 miles) from the well.

damaged, and net changes to subsurface pore pressure caused by the energy project. These net changes involve the volume and pressure of fluids injected or extracted, the duration of injection and extraction, and the number of wells involved in the project. Note that these variables may be related; that is, the total fluid volume depends on the duration of injection or extraction and the number of wells involved.

Two spatial aspects of risk analysis are important to consider in the context of induced seismicity:

1. **Multiple structures that can be damaged**. A single well that induces earthquakes large enough to cause damage at the surface may damage multiple structures at the surface. If seismicity migrates during well operations (which is common for disposal wells), earthquakes have multiple opportunities to impact many structures. Even a small community located near a single well will have multiple structures with a range of vulnerabilities to ground shaking. Multiple structures give an increased chance of having one or a few structures with very weak resistance to ground shaking. Operations located in areas with many structures, such as the Basel, Switzerland, geothermal project, clearly have higher risk than a similar project in an unpopulated area. Likewise, CCS operations that are located at power plants in or near urban areas and which have the potential through injections of large amounts of CO_2 over long time periods to increase reservoir pressures over large areas that may have surface developments may have increased risk.
2. **Multiple well locations**. The risk associated with induced seismicity has to be evaluated in terms of the sources of human activities. A geothermal operation, for example, may have multiple injection wells, each of which may generate seismic events that can affect different communities. For a large petroleum field, multiple wells may be used to inject fluid for secondary recovery, and each well may generate earthquakes that can affect separate communities. The spatial distribution for an entire industry project (e.g., underground injection of CO_2) may be very large, and a risk analysis of the entire project would necessarily include that large spatial distribution and the multiple structures in that spatial area which induced seismic events might affect.

If a small number of wells (e.g., 10) are put in operation, the maximum shaking associated with earthquakes induced by those 10 wells can be described (Figure 5.4). In this example, a majority of wells (9 out of 10) will produce only felt motion, and only 1 out of 10 will produce ground motion with the potential for minor damage. No observations of moderate or greater (abbreviated hereafter as "moderate+") damage occur in this example.

If many wells (e.g., 1,000) are put into operation, a histogram of the maximum shaking induced by those 1,000 wells would show that 250 wells are expected to produce ground motions capable of minor damage to structures. Ten wells are expected to produce ground motions capable of moderate+ damage to structures.

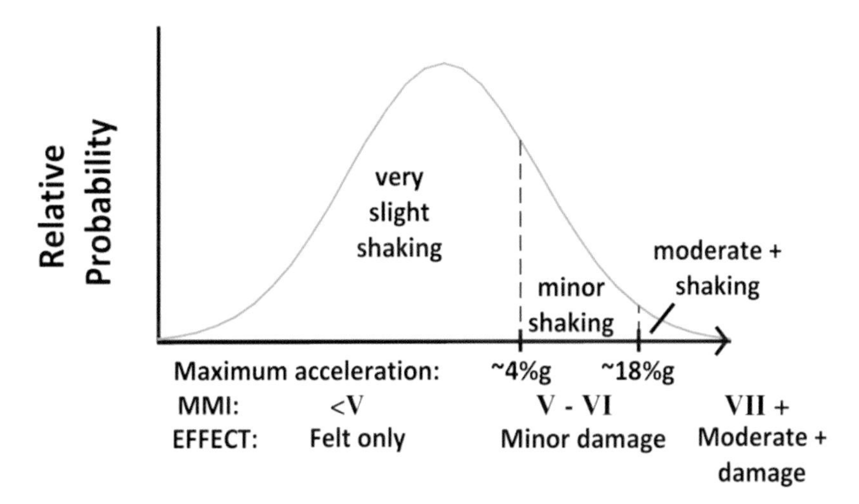

FIGURE 5.4 Example of relative probability distribution of maximum shaking at the ground surface from induced seismicity caused by one well. The relative probability increases upward on the vertical axis. The horizontal axis shows several kinds of measurements or effects of ground shaking: the upper scale indicates the amount of shaking (slight through moderate+); the second scale indicates ground acceleration, which increases from left to right; the next scale indicates MMI or the Modified Mercalli Intensity scale, which indicates the level of ground shaking at a particular location and has units designated by Roman numerals, also increasing from left to right in the level of ground shaking (see also Chapter 1); and the lower scale is the "felt" effect, ranging from "felt only" on the left through minor to moderate or greater ("moderate+") damage. The probability of very slight shaking is much higher than for moderate+ shaking (or damage) for one well that causes an induced seismic event of any magnitude.

A more general distribution of ground motion from a range of earthquakes with ground motions quantified by the largest horizontal acceleration[1] that occurs shows that the majority of shaking will be in the category of "felt only" (Figure 5.4). A small percentage (~25 percent) may have the potential to cause minor damage, and a very small percentage (~1 percent) may have the potential to cause moderate+ damage (Figure 5.4).

The important conclusion is that, while the risk of minor, moderate, or heavy damage from induced earthquake shaking may be small for each individual well, a large, spatially distributed operation leads to a higher probability of such damage. If we define P_M as the probability of moderate+ damage given surface ground motion from one well, then the prob-

[1] The peak horizontal acceleration of the ground is a common measure of ground shaking because the maximum force on objects sitting on the ground is proportional to the peak horizontal acceleration through Newton's second law. Acceleration is measured in units of gravity, "*g*," which is the acceleration of a falling object. For comparative purposes, a modern, high-powered sports car can accelerate at about 50%g.

TABLE 5.1 Probability of Damage Increases with Number of Wells

Total Number of Wells (N)	$[P_M]_{N\,wells}$	Expected Number of Wells Causing Moderate+ Damage
1	1%	0
5	5%	0
10	10%	0
100	63%	1
1,000	99.9%	10

ability of at least one observation of moderate+ damage given that N wells are in operation can be calculated[2] as

$$[P_M]_{N\,wells} = \text{probability of 1 or more moderate+ damaging motions for } N \text{ wells} = 1 - (1 - P_M)^N$$

This probability increases with the number of wells N (for $P_M = 1\%$), as shown in Table 5.1.

This example illustrates that, as an industry begins operation with a few wells, there might be no apparent problem with induced seismicity. As the industry expands to 100, 1,000, or more wells, there can be a significant likelihood that induced seismicity will cause damage to structures somewhere, as a result of the large number of earthquakes and ground motions that are induced, even though the probability of any one well producing such ground motions is small.

Tectonic earthquakes cause some level of earthquake risk for buildings, primarily in areas like California with relatively frequent events. Seismic building codes provide some level of protection but are not a guarantee against earthquake damage. In other regions, building codes provide lower levels of seismic protection, and earthquakes (whether tectonic or induced) may cause damage, depending on the level of ground motion associated with them.

QUANTIFYING HAZARD AND RISK

Several steps can be taken to quantify hazard and risk. As described in the previous section, the quantification of hazard and risk requires probability assessments, which may be

[2] This is a special case of the Bernoulli distribution with N independent trials and probability P_M of occurrence of the phenomenon of interest (moderate+ damage). The probability of at least one observation of this damage is 1 minus the probability of no observations of this damage, given N independent trials. Any dependence among ground motions for a given technology can be examined as part of the hazard assessment step identified in the section Quantifying Hazard and Risk (this chapter), in particular step 3 in Table 5.2.

either statistical (based on data) or analytical (based on scientific and engineering models). Thus, for implementation, some of the steps will require the collection of statistical data. Other steps can modify and use analytical models that have been developed for hazard and risk analysis of tectonic earthquakes. Table 5.2 summarizes the steps that can be taken to quantify the hazard and risk of induced seismicity for a single project (a single wastewater disposal well, oil or gas extraction well, etc.).

Step 1 in Table 5.2 involves estimating the probability of generating earthquakes with $\mathbf{M} \geq 2.0$. This is a statistical problem that can be addressed only by collecting statistical data on the number of wells drilled for each technology, their characteristics (depth, volumes of fluids, pressures, rates of injection or extraction), and observations on whether they generate earthquakes. Simulation models that predict fluid flow in the Earth's crust given characteristics such as permeability, pumping rate and volume versus time, geologic units (including ages of the rocks), and other factors can be the basis for predictive models, and these models can be refined on a probabilistic basis as more data and observations are gathered and analyzed. The cell labeled "1C" in Table 5.2 indicates that these statistics will be technology dependent, because the typical volume of fluid, pressure at which it is injected, and other factors in a given project depend on the energy technology. Cell 1D indicates that energy projects in tectonically active regions can be expected to have a higher probability of generating $\mathbf{M} \geq 2.0$ earthquakes than do energy projects in tectonically stable regions. Finally, cell 1E indicates that the probability assessment from statistics will have a depth dependence: large earthquakes are less likely to be induced by shallow wells.

Step 2 involves estimating the probability of felt shaking at the surface (see cell 2A, Table 5.2). This is an analytical problem with some statistical inputs (cell 2B). Specifically, data are needed on the frequencies of occurrence of different earthquake magnitudes. As cell 2C indicates, these frequencies are expected to be technology dependent. The reason is that, among energy technologies, earthquake-generation mechanisms vary (Chapter 2), and the net injected or extracted fluid volume varies (Chapter 3). Once these data on magnitude distributions are obtained, analytical methods are available to estimate shaking (see, for example, Boore, 2003). This probability may be region dependent (cell 2D) because earthquakes in stable crustal regions may release higher levels of crustal stress than similar-magnitude events in active crustal regions. Finally, the probability of felt shaking will depend on the depth of the induced earthquakes (cell 2E) (see Figures 5.2 and 5.3).

Step 3 involves estimating the probability of different strengths of earthquake shaking (cell 3A). This is a well-studied problem in seismic hazard analysis for tectonic earthquakes, for which analytical techniques are available (cell 3B). The result will depend on energy technology (cell 3C) because observations of earthquake magnitude distributions, particularly the maximum magnitude, have some dependence on energy technology (see Figure 3.15). Also, the result will depend on region (cell 3D) and depth (cell 3E), because earthquake magnitude distributions depend on these factors.

TABLE 5.2 Steps for Hazard and Risk Assessment for a Single Project

Step (see corresponding box in Figure 5.1)	A. Probability needed	B. Method	C. Technology Dependent?	D. Region Dependent?	E. Depth Dependent?
1	1A. P[generate **M** ≥ 2 earthquakes]	1B. Statistical	1C. Yes, depends on factors such as volume, pressure, rate, and depth	1D. Yes, tectonically active versus stable region	1E. Yes, large earthquakes usually not induced near surface
2	2A. P[shaking felt at surface]	2B. Analytical/ Statistical	2C. Yes, depends on magnitude distribution and maximum magnitude	2D. Yes, depends on earthquake properties	2E. Yes, deeper induced earthquakes may not be felt
3	3A. P[strength of shaking]	3B. Analytical	3C. Yes, depends on maximum magnitude	3D. Yes, depends on earthquake properties	3E. Yes, shallow earthquakes will generate stronger shaking
4	4A. P[structures and people affected]	4B. Analytical	4C. No	4D. Yes, depends on structural strength and tolerance for shaking	4E. Yes, deeper earthquakes, if felt at the surface, may affect a larger area

NOTE: Gray shaded cells indicate methods that have to be developed to estimate probabilities ("P") for various aspects of an induced seismic event shown in the green-shaded cells. These four aspects include the probability of generating an earthquake of **M** > 2.0, the probability of shaking being felt at the surface, the probability of different strengths of shaking from an earthquake, and the probability that the earthquake shaking will affect structures and people.

Finally, step 4 involves estimating the probability that structures and people are affected (cell 4A). Analytical methods for seismic risk analysis (cell 4B) are well established for tectonic earthquakes, and these should be applicable to induced earthquakes. The methods will *not* depend on technology (cell 4C), because a structure's response does not depend on how the shaking was generated. However, the methods do depend on region (cell 4D); structures outside of California and Alaska are generally not designed to withstand high levels of ground shaking, and people in aseismic regions may be less tolerant of low-level shaking than those who have previously felt natural earthquakes. Deeper earthquakes will have an influence on the numbers of structures and people affected (cell 4E) if the associated earthquake shaking covers a wide region and affects more structures and people.

Table 5.2 summarizes steps that can be taken to estimate hazard and risk for individual energy projects. The specific statistical data that need to be collected, and analytical methods that need to be modified from other fields, are summarized in column B. Each of the statistical or analytical methods in column B will calculate the probability indicated in the corresponding cell in column A, and these calculations will depend on the corresponding cells in columns C, D, and E. For instance, statistical data on $\mathbf{M} \geq 2$ earthquake generation (cell 1B) need to be collected and analyzed by energy technology, volume of fluid, injection pressure, rate of injection, etc. An unstated assumption in Table 5.2 is that data are to be collected for *new energy projects* in areas that are known to have a history of induced seismicity, as well as existing projects. The reason is that, going forward, we presumably are interested in estimating hazard and risk from induced seismicity caused by further expansion of energy production, not by existing energy production. However, data from existing projects will allow forecasts of induced seismicity for industries as a whole. The distinction is important: seismicity induced by a new injection or disposal well will differ from seismicity induced by a well that has been in production for years, where crustal stresses may have equilibrated.

Note that steps 1 through 3 apply regardless of whether the potential induced seismicity will occur in areas of high population or sparse population. Step 4 determines the effect on structures and people, and this effect of course depends on the location with respect to structures at risk and people. Induced seismicity could be caused in a region of sparse population, affecting few people, but could affect dams, bridges, or power plants, with large concurrent costs.

These steps, if developed, can be used in three important ways:

First, by compiling statistics on earthquake generation by technology and characteristics (cell 1C), insight can be gained on what combinations of volumes, pressures, rates of injection/extraction, and so on lead to higher probabilities of induced seismicity. This insight can be used to create well-documented, data-based input to best practices protocols (see also Chapter 6).

Second, energy technology development, whether through public or private efforts, will have data with which to make decisions to minimize induced seismicity effects on people and structures. For example, if a particular project is observed to generate **M** ≥ 2 earthquakes (i.e., the probability in cell 1A becomes 1 for that project), decisions can be made on pumping characteristics to minimize the probabilities of shaking felt at the surface (cell 2A) and of strong shaking (cell 3A).

Third, the calculated probabilities of shaking felt at the surface (cell 2A), of strong shaking (cell 3A), and of structures and people being affected (cell 4A) can be generalized from those for one project (as depicted in Table 5.2) to forecast the total number of induced seismicity cases that will occur and the number of structures and people affected. If detailed statistical data can be obtained for cells 1B and 2B, this generalization can account for details on forecast locations of projects, volumes and other characteristics of pumping, and proximity to inhabited areas. The estimated numbers of people and structures affected can then become the basis for decisions on whether and how to minimize the impacts of induced seismicity.

Directed research could support development of these steps for the quantification of hazard and risk, with the overall goal of integrating these steps to improve our capability to predict induced events and their consequences. Chapter 6 develops these ideas further by discussing best practices and protocols to avoid or mitigate the impacts of induced seismicity during energy development projects.

REFERENCES

BGS (British Geological Survey). 2011. Blackpool earthquake, Magnitude 1.5, 27 May 2011. Available at www.bgs.ac.uk/research/earthquakes/blackpoolMay2011.html (accessed November 2011).

Boore, D.M. 2003. Simulation of ground motion using the stochastic method. *Pure and Applied Geophysics* 160:635-676.

CHAPTER SIX

Steps Toward a "Best Practices" Protocol

THE IMPORTANCE OF CONSIDERING THE ADOPTION OF BEST PRACTICES

This report has shown that induced seismicity may be associated with the development of different energy technologies involving fluid injection and sometimes fluid withdrawal (see, e.g., Chapter 3). Furthermore, despite an increased understanding of the basic causes of induced seismicity (Chapter 2), these kinds of energy development projects will retain a certain level of risk for inducing seismic events that will be felt by members of the public (see Chapter 5). While the events themselves are not likely to be very large or result in any significant damage, they will be of concern to the affected communities and thus require both attention before an energy project involving fluid injection gets under way in areas of known seismic activity (whether tectonic or induced) and management and mitigation of the effects of any felt seismic events that occur during operation.

This chapter outlines specific practices that consider induced seismicity both before and during the actual operation of an energy project and that could be employed in the development of a "best practices" protocol specific to each energy technology. The aim of any eventual best practices protocol would be to diminish the possibility of a felt seismic event from occurring, and to mitigate the effects of an event if one should occur. The committee views the ultimate successes of any such protocol as being fundamentally tied to the strength of the collaborative relationships and dialogue among operators, regulators, the research community, and the public (see also Chapter 4). Indeed, protocols, when properly developed and understood, can serve to protect and benefit the various parties involved both directly and indirectly in energy project development.

The chapter begins with a few examples of induced seismicity "checklists" and protocols in the literature that have been developed for the purpose of management of induced seismicity for specific energy projects. The chapter then discusses some of the key components of these checklists and protocols and develops two induced seismicity protocol "templates," one for enhanced geothermal systems and another for wastewater injection wells. The chapter includes discussion of the incorporation of a "traffic light" system to manage fluid injection and concludes with a discussion of the role and importance of public outreach and engagement prior to and during development of energy projects involving fluid injection. The committee acknowledges that this kind of preemptive management approach embodied in any best practices protocol for induced seismicity can be complicated by the challenges of determining whether any seismicity felt in a region with injection wells is induced or is due to natural, geologic causes (see Chapter 1). However, we suggest that the benefit of the collective dialogue and establishing best practices in the event of a felt seismic event is in itself constructive, with few or no negative consequences.

EXISTING INDUCED SEISMICITY CHECKLISTS AND PROTOCOLS

Induced seismicity does not fall squarely in the sole purview of any single government agency and, in fact, requires input and cooperation among several local, state, and federal entities, as well as operators, researchers, and the public (see Chapter 4). Because of these shared interests and potential responsibilities, the committee suggests that the agency with authority to issue a new injection permit or the authority to revise an existing injection permit is the most appropriate agency to oversee decisions made with respect to induced seismic events, whether before, during, or after an event has occurred. In many cases this responsibility would fall to state agencies that permit injection wells. In areas that are known by experience to be susceptible to induced seismicity, a best practices protocol could be incorporated into the approval process for any proposed (new) injection permit. In areas where induced seismicity occurs, but was not anticipated in a particular area, existing injection permits relevant to that area could be revised to include a best practices protocol.

Two Checklists to Evaluate the Potential for Induced Seismicity and the Probable Cause of Observed Events

Checklists can be convenient tools for government authorities and operators to discuss and assess the potential to trigger seismic events through injection, and to aid in determining if a seismic event is or was induced. Two checklists, one to address each of these two circumstances—the potential for induced seismicity and the determination of the cause of a felt event—were developed nearly two decades ago by Davis and Frohlich (1993) to address each of these circumstances (summarized in the sections that follow). Their work recommends a list of ten "yes" or "no" questions to quantify "whether a proposed injection project is likely to induce a nearby earthquake" and a list of seven similar questions to quantify "whether an ongoing injection project has induced an earthquake."

Will Injection Induce Earthquakes: Ten-Point Checklist

The ten-question checklist evaluates four factors related to possible earthquake hazards: historical background seismicity, local geology, the regional state of stress, and the nature of the proposed injection. Table 6.1, modified from Davis and Frohlich (1993), compares

TABLE 6.1 Criteria to Determine if Injection May Cause Seismicity

	Question	NO APPARENT RISK	CLEAR RISK	Texas City, Texas	Tracy, Quebec	Denver RMA, Colorado
	Background Seismicity					
1a	Are large earthquakes (**M** $\geq$ 5.5) known in the region (within several hundred km)?	NO	YES	NO	YES	YES
1b	Are earthquakes known near the injection site (within 20 km)	NO	YES	NO	YES	NO?
1c	Is rate of activity near the injection site (within 20 km) high?	NO	YES	NO	NO	NO
	Local Geology					
2a	Are faults mapped within 20 km of the site?	NO	YES	YES	YES	NO?
2b	If so, are these faults known to be active?	NO	YES	NO	NO	NO
2c	Is the site near (within several hundred km of) tectonically active features?	NO	YES	NO?	YES	YES
	State of Stress					
3	Do stress measurements in the region suggest rock is close to failure?	NO	YES	NO	NO?	YES[a]
	Injection Practices					
4a	Are (proposed) injection practices sufficient for failure?	NO	YES	NO?	YES	YES[a]
4b	If injection has been ongoing at the site, is injection correlated with the occurrence of earthquakes?	NO	YES	NO	N.A.	N.A.
4c	Are nearby injection wells associated with earthquakes?	NO	YES	NO	N.A.	N.A.
	TOTAL "YES" ANSWERS	0	10	**1**	**5**	**4**

[a] Assumes stress measurements completed prior to survey.
NOTE: RMA, Rocky Mountain Arsenal.
SOURCE: Davis and Frohlich (1993).

the answers of this ten-point criteria list for three injection wells. The wells listed include an existing injection well located in Texas, a proposed injection project in Quebec, and the injection well located at Rocky Mountain Arsenal in Denver with questions answered "as if injection had not yet taken place."

The authors note, "In actuality, if one were to propose injection at a site near Denver today, the existence of the earthquake activity between 1962 and 1972 would alter the profile, and there would be six or more 'yes' answers" (p. 214). The authors go on to say, "At the Tracy, Quebec site we find five 'yes' answers. . . . We would thus conclude that the situation is more similar to Denver than the Texas Gulf Coast" (p. 214).

Did Injection Induce the Observed Earthquake(s): Seven-Point Checklist

The list of seven questions from Davis and Frohlich (1993) again evaluates four factors related to possible cause: background seismicity, temporal correlation, spatial correlation, and injection practices. In Table 6.2 the seven questions are listed and are specifically phrased so that a "yes" answer would indicate underground injection induced the earthquake(s) and a "no" answer would indicate the earthquake(s) were not caused by injection.

Two injection wells are evaluated in Table 6.2. The well in Denver, Colorado, was the injection well at the Rocky Mountain Arsenal, which was definitely shown to be the cause of induced earthquakes in the mid-1960s. The Painesville, Ohio, well, also known as the Calhio well, which was injecting liquid waste from agricultural manufacturing, was investigated as a cause of earthquakes and revealed ambiguous results; the scientists who examined the data could not make a certain correlation between the injection well and the earthquakes, in part due to historical (natural) seismic activity in the area.[1]

An Example Best Practices Protocol for Induced Seismicity Associated with Enhanced Geothermal Systems

As an example of a protocol used in projects expected to result in induced seismicity, the Department of Energy (DOE) has published a best practices protocol for addressing the potential of induced seismicity associated with the development of enhanced geothermal systems (EGS) (Majer et al., 2012). The steps that a developer might follow in that protocol are summarized in Box 6.1. The DOE states that this protocol is not intended as a proposed substitute to existing local, state, and /or federal regulations but instead is intended to serve as a guideline for the systematic evaluation and management of the anticipated effects of the induced seismicity that are expected to become related to the development of an EGS project.

[1] For example, see www.dnr.state.oh.us/geosurvey/earthquakes/860131/860131/tabid/8365/Default.aspx.

TABLE 6.2 Seven Questions Forming a Profile of a Seismic Sequence

	Question	Earthquakes Clearly NOT Induced	Earthquakes Clearly Induced	I Denver, Colorado	II Painesville, Ohio
	Background Seismicity				
1	Are these events the first known earthquakes of this character in the region?	NO	YES	YES	NO
	Temporal Correlation				
2	Is there a clear correlation between injection and seismicity?	NO	YES	YES	NO
	Spatial Correlation				
3a	Are epicenters near wells (within 5 km)?	NO	YES	YES	YES?
3b	Do some earthquakes occur at or near injection depths?	NO	YES	YES	YES?
3c	If not, are there known geologic structures that may channel flow to sites of earthquakes?	NO	YES	NO?	NO?
	Injection Practices				
4a	Are changes in fluid pressure at well bottoms sufficient to encourage seismicity?	NO	YES	YES	YES
4b	Are changes in fluid pressure at hypocentral locations sufficient to encourage seismicity?	NO	YES	YES?	NO?
	TOTAL "YES" ANSWERS	0	7	6	3

SOURCE: Davis and Frohlich (1993).

Using this protocol as a foundation, the committee has adapted the protocol's set of seven steps in Table 6.3 to illustrate a set of parallel activities, with steps 2 through 7 undertaken essentially concurrently, as opposed to sequentially, to help manage and mitigate induced seismicity from injection associated with EGS. Viewing a protocol as a set of parallel activities is useful not only for general project management but also for the ability it provides to reassess the protocol through time as circumstances of an energy project change and more data are acquired. This resulting matrix form can be used as a template to develop an appropriate protocol to mitigate the potential to induce seismicity in other

BOX 6.1
The Department of Energy Protocol for Addressing Induced Seismicity Associated with Enhanced Geothermal Systems

The elevated downhole fluid pressures used in EGS induce fracturing that can result in a level of induced seismicity that is felt at the surface and that in some cases has caused serious concern among those living nearby (see Chapter 3). To attempt to avoid the repeated occurrence of such results, while encouraging the future use of geothermal resources, a protocol has evolved to serve as a guide for EGS developers within the United States as well as internationally. The most current protocol, developed by the Department of Energy (Majer et al., 2012), "outlines the suggested steps that a developer should follow to address induced seismicity issues, implement an outreach campaign and cooperate with regulatory authorities and local groups." This sequence of seven steps can be summarized as follows:

STEP 1. *Perform Preliminary Screening Evaluation.* Assess the feasibility of the proposed project as to its technical, socioeconomic, and financial risks in order to provide an initial measure of the project's potential acceptability and ultimate success. Review local regulatory conditions, the level of natural seismicity, and the probable impacts of the project on any nearby communities and sensitive facilities.

STEP 2. *Implement an Outreach and Communication Program.* Before operations begin, implement a public relations plan that describes the proposed operations, determine the resulting concerns, address those concerns, and then periodically meet with the locals to explain the upcoming operations and the results of the work done to date.

STEP 3. *Review and Select Criteria for Ground Vibration and Noise.* Identify and evaluate local environmental and regulatory standards for induced vibration and noise. Develop appropriate acceptance criteria for an EGS project.

STEP 4. *Establish Local Seismic Monitoring.* Collect baseline data on the regional seismicity that exists before operations begin. Install and operate a local seismometer array to monitor the project's operations.

STEP 5. *Quantify the Hazard from Natural and Induced Seismic Events.* Estimate the ground shaking hazard from the natural seismicity to provide a baseline to evaluate the additional hazard from the induced seismicity.

STEP 6. *Characterize the Risk of Induced Seismic Events.* Characterize the expected induced ground motion and identify the assets and their vulnerability within the area likely to be influenced by the project.

STEP 7. *Develop a Risk-Based Mitigation Plan.* If the level of seismic impacts becomes unacceptable, direct mitigation measures are needed to further control the seismicity. A "traffic light" system can allow operations to continue as is (GREEN), or require changes in the operations to reduce the seismic impact (AMBER), or require a suspension of operations (RED) to allow time for further analysis. Indirect mitigation may include community support and compensation.

energy technologies. The committee has done this exercise for induced seismicity associated with injection wells used for oil and gas development (Environmental Protection Agency [EPA] Underground Injection Control [UIC] Class II wells) or with carbon storage (EPA UIC Class VI wells) and has developed an example of the primary elements that might be included in a best practices protocol matrix (Table 6.4).

THE USE OF A TRAFFIC LIGHT CONTROL SYSTEM

The protocols described in Box 6.1 and Tables 6.3 and 6.4 refer to a "traffic light" control system for responding to an instance of induced seismicity. Such a system, although rarely employed in energy technology projects with active cases of induced seismicity,[2] allows for low levels of seismicity but adds additional monitoring and mitigation requirements when seismic events are of sufficient intensity to result in a concern for public health and safety. The preferred criterion to be used for such a control system has been the level of ground motion observed at the site of the sensitive receptor, be it a public or private facility. Seismic event magnitude alone is generally insufficient as the only criterion because of the nature of attenuation (absorption or loss of energy) with increasing distance from an event location to a sensitive receptor site. Zoback (2012) provides a summary of a traffic light system for the purpose of managing potential induced seismicity from wastewater disposal.

As an example, the Bureau of Land Management (BLM) recently issued as its "Conditions of Approval"[3] for a proposed EGS project the specific procedures to be followed in the event that induced seismicity is observed to be caused by the proposed stimulation (hydraulic fracturing) operation. The specific procedures included the use of the traffic light control system that allows hydraulic fracturing to proceed as planned (green light) if it does not result in an intensity of ground motion in excess of Mercalli IV ("light" shaking with an acceleration of less than 3.9%g), as recorded by an instrument located at the site of public concern. However, if ground motion accelerations in the range of 3.9%g to 9.2%g are repeatedly recorded within one week, equivalent to Mercalli V ("moderate" shaking), then the operation is required to be scaled back (yellow light) to reduce the potential for the further occurrence of such events. And finally, if the operation results in a recorded acceleration of greater than 9.2%g, resulting in "strong" Mercalli VI or greater shaking, then the active operation is to immediately cease (red light).

The authority for the permitting of Class II injection well location varies by state and is discussed in Chapter 4. Well permits of Class II injection wells in Colorado, for example, are reviewed by the Colorado Geological Survey (COGCC, 2011). During a geologic review,

[2] To the committee's knowledge, the traffic light system has been applied only at the Berlin geothermal field in El Salvador (Majer et al., 2007) and at Basel, Switzerland.

[3] R.M. Estabrook, BLM, Conditions of Approval for GSN-340-09-06, Work Authorized: Hydroshear, The Geysers, January 31, 2012.

TABLE 6.3 Primary Elements of a Protocol for Addressing Induced Seismicity in EGS Technologies Adapted as a Series of Parallel Activities Extending over the Lifetime of the Operation

Initial Screening to Determine the Feasibility of the EGS Project	Assess the local hazard potential from natural seismicity; the local, state, and federal regulations; the nearness of the project to population centers; the probable magnitude of induced events; and the probable risks of potential damage from both natural and induced events. If the proposed EGS project appears to be feasible based on this initial screening assessment, then the **Essential Activities of the EGS project as listed below** are recommended to proceed in the manner described within each of the five sequential stages of project development as identified herein.				
Category of Essential Activities	PREPARATION STAGE	DRILLING STAGE	STIMULATION STAGE	OPERATIONS STAGE	COMPLETION STAGE
Public and Regulatory Communications	Identify the local people and organizations to be met with. Hold an initial public meeting, explain the planned project, identify their concerns.	Meet with and inform the public, regulators, and media as to the drilling schedule. Upon completion meet and explain the drilling results.	Meet with and inform the public, regulators, and media as to the stimulation schedule and results.	Meet with and inform the public, regulators, and media as to the operations schedule and results.	Meet with and inform the public, regulators, and media as to the project completion.
Criteria for Ground Vibration and Noise	Install ground motion and noise monitoring instrumentations.	Report to the public, regulators, and media the monitoring results.	Report to the public, regulators, and media the monitoring results.	Report to the public, regulators, and media the monitoring results.	Report to the public, regulators, and media the monitoring results.
Seismic Monitoring	Determine areal size and sensitivity needed for local array. Install and operate the seismic recording array and allow timely public access to results.	Continue to monitor the seismicity recorded and publically report the results.	Add and/or reposition array's seismometers as needed to follow and characterize the induced events.	Add and/or reposition array's seismometers as needed to follow and characterize the induced events.	Continue to record and report on the induced seismicity as long as needed to describe the local conditions.

Hazard Assessment	Evaluate the potential additional hazard to be expected from the locally induced seismicity.	Review and reassess the potential for damage based on local observations.	Review and reassess the potential for damage based on local observations.	Review and reassess the potential for damage based on local observations.	Report to the public, regulators, and media on any actual hazards observed.
Risk Assessment	Develop a probabilistic risk analysis to estimate the probability of risk (monetary loss) to be expected.	Revise the Risk Assessment as appropriate, based on any physical damage, nuisance, and/or economic losses attributed to the project operations.			Report to the public, regulators, and media on the actual results experienced.
Direct Mitigation Plans	Develop a plan to control the level and impact of locally induced seismicity.	If needed, implement the control system to cause the drilling, stimulation, or continuing operations to be temporarily reduced or suspended until the level of the locally induced seismicity has been returned to an acceptable level, as determined by the regulatory agencies.			Report to the public, regulators, and media on the actual results experienced.
Indirect Mitigation Plans	Provide local jobs, support local community facilities, and provide compensation if appropriate. Continue indirect mitigation activities as long as needed.				

TABLE 6.4 Summary of the Primary Elements of a Protocol for Addressing Induced Seismicity Associated with Injection Wells Used for Oil and Gas Development (EPA UIC Class II wells) or Associated with Carbon Sequestration (EPA UIC Class VI wells)

	Additional UIC Permitting Requirements	After Drilling and Prior to Injection (A Second Look)	Monitoring Requirements During Injection
Public and Regulatory Communications	Operator should identify local residents and cities and counties that could be affected by induced seismicity and hold public meetings to explain project and identify concerns.	Operator should notify appropriate regulatory agencies and the local public and provide updated information and analysis based on any new information obtained during drilling operations.	Operator should provide periodic updates to appropriate regulatory agencies and the local public on the locations and extent of their injection operations and the locally observed seismic activity.
Hazard Assessment	Evaluate the potential additional hazard to be expected from locally induced seismicity.	Review and reassess the potential for induced seismicity based on any additional information obtained during drilling and completion of the injection well.	Report to the appropriate regulatory agencies and the public on any actual hazards observed during injection activity.
Risk Assessment	Develop a probabilistic risk analysis to estimate the probability of risk to be expected.	Revise the risk assessment as appropriate based on any additional information obtained during the drilling and completion of the injection well.	Revise the risk assessment as appropriate based on additional information obtained during injection activity.
Criteria for Ground Vibration	Determine areal size, sensitivity, and appropriate instrumentation needed for local array.		

Seismic Monitoring	Install and operate the seismic recording array to obtain baseline seismic data and record seismic events due to injection activity.		
Mitigation Plans	Develop a plan to control the level and impact of locally induced seismicity based on the hazard and risk assessment and baseline seismic data.	Revise mitigation plan as appropriate based on any additional information obtained during the drilling and completion of the injection well.	Continuously review and assess mitigation plan to determine effectiveness.

NOTE: The entire protocol would apply to injection wells proposed in areas where induced seismicity has actually occurred. In areas where induced seismicity was not expected but later occurred, the shaded requirements would apply as revisions to the original injection permit.

the historical earthquake data near the well are closely examined, along with any published fault maps in the area. Additional data regarding fault information, such as that available from three-dimensional (3D) seismic images or other geological information from the well operator may be requested if the well appears to be sited in a high-risk area.

MITIGATING THE EFFECTS OF INDUCED SEISMICITY ON PUBLIC AND PRIVATE FACILITIES

The best practices protocols appropriately include an emphasis on establishing a public relations plan to inform the public as well as the appropriate regulatory agencies of the purpose of the proposed or existing project, the intended operations, and the expected impacts on the nearby communities and/or facilities. Public acceptance begins with an understanding of what is expected to transpire and what contingencies exist for dealing with the unexpected. Inherent in any public information and communication plan is the idea that a developer regularly meets with the local public to explain the schedule and activities of each upcoming stage of operations, as well as the results of the operations performed to date. During the committee's information gathering session in The Geysers in Northern California and at the associated workshop in Berkeley, we had an opportunity to discuss the 50-year history of induced seismicity at The Geysers geothermal field and meet with the operators, regulatory authorities, researchers, and the local residents from Anderson Springs and Cobb, nearest to The Geysers operations, and subject to the effects of ground shaking due to induced seismicity (see Appendix B—meeting agenda). The discussions we had with these individuals provided some interesting lessons (Box 6.2) regarding the value and potential success of constructive public engagement, for all parties, when induced seismicity may be or becomes an issue in an energy development project. The committee found several very important points to consider regarding the value of successful public outreach, using this example from The Geysers:

1. **Time**. Public engagement, even if begun early in a project's planning processes, is a process that occurs over a long time and not a goal in itself. As a process, public engagement requires dedicated and frequent communications among industry, the public, government officials, and researchers.
2. **Information and education**. Although the initial burden to supply information and to educate local residents lies with the operator and government authorities, residents, too, have a responsibility to become informed and to be constructive purveyors of data and information back to those responsible for operations to allow constructive dialogue to take place.
3. **Managed expectations through transparency**. Coupled to the sharing of information and education is the idea of managing expectations. Each group involved

BOX 6.2
The Geysers: Toward Mitigating the Effects of Induced Seismicity

About 40 years ago researchers at the U.S. Geological Survey (USGS) and elsewhere began reporting that induced seismicity was associated with the geothermal production and injection operation at The Geysers (e.g., Hamilton and Muffler, 1972). At first, the causes of the seismicity in this area, where natural seismic activity has a long history, were unclear to the seismologists and to the local operators. Following the installation of additional seismometers to increase the accuracy of locating the events, it became evident that the earthquakes were primarily associated with the injection wells associated with The Geysers and, indeed, essential for continued operation of the field to produce electricity (see Chapter 3; Box 3.1). Consequently, when a pipeline project was proposed 15 years ago to deliver wastewater for increased injection at The Geysers to maintain and enhance power generation, the Environmental Impact Report required the establishment of a Seismic Monitoring Advisory Committee (SMAC) to monitor and report on the production and injection, and seismic activities.

The committee includes representatives of the Bureau of Land Management and California state regulatory agencies, county government, the USGS and Lawrence Berkeley National Laboratory, the local communities, and the operators of the geothermal facilities. Real-time results of the seismic monitoring are continuously available to all at the Northern California Seismic website, and the semiannual meetings of this committee provide a forum for all the stakeholders to compare the locations and magnitudes of the reported seismic events to the locations of the reported production and injection activities.

Despite the benefits of establishing the SMAC, the geothermal operators were still viewed by some local residents as not having taken sufficient responsibility for mitigating the effects of the clearly increased numbers of induced seismic events being felt within the local communities (see Box 3.1), and a petition was filed to declare the situation as being a public nuisance. The county government established two subcommittees to deal directly with the residents of the two local communities of Anderson Springs and Cobb. Each subcommittee has representatives of its local community, the local operators, and the local county supervisor. Ground motion recording instruments were installed in each community, and the resulting information is available in near real time at an independently controlled website. This information allows anyone with Internet access to compare the recorded time of an observed ground motion with the reported times of the separately reported local seismic events in order to determine the location of the apparent source that caused the observed ground motion.

The members of each subcommittee have developed a system of receiving, reviewing, and approving damage claims attributed to the local induced seismicity. Over the past 6 years the geothermal operators have reimbursed the homeowners for their costs to have their home damages repaired, at a total expense of less than $100,000 while contributing funds far in excess of this for improvements to the common facilities in the local communities. In addition the county government has continued to contribute to these communities part of the mitigation funds it receives as redistributions of the royalty payments made to the federal government by the local geothermal operators. This system of coordinating the use of the combined resources of both industry and local government has much improved the mitigation of the effects of the locally induced seismicity, and it is now resulting in much improved and mutually satisfactory relationships among the parties.

SOURCES: DOE (2009); J. Gospe, Anderson Springs Community Alliance, 2011, "Man-Made Earthquakes & Anderson Springs," DVD, June 30; see also www.andersonsprings.org/.

in an energy development project has different goals and expectations. Mutual understanding of other groups' goals and expectations is fundamental to developing strong and constructive communication. Transparency regarding these goals and expectations is important to their management.

REFERENCES

COGCC (Colorado Oil and Gas Conservation Commission). 2011. COGCC Underground Injection Control and Seismicity in Colorado. January 19. Denver, CO: Department of Natural Resources. Available at cogcc.state.co.us/Library/InducedSeismicityReview.pdf (accessed February 2012).

Davis, S.D., and C. Frohlich. 1993. Did (or will) fluid injection cause earthquakes? *Seismological Research Letters* 64(3-4):207-224.

DOE (U.S. Department of Energy). 2009. Appendix J—Statement of Compliance with DOE Seismicity Protocol. Geysers Power Company's Enhanced Geothermal System Demonstration Project, Northwest Geysers Geothermal Field, Sonoma County, California (DOE/EA 1733). Available at www.eere.energy.gov/golden/NEPA_FEA_FONSI.aspx (accessed February 2012).

Hamilton, R.M., and L.J.P. Muffler. 1972. Microearthquakes at The Geysers geothermal area, California. *Journal of Geophysical Research* 77:2081-2086.

Majer, E.L., R. Baria, M. Stark, S. Oates, J. Bonner, B. Smith, and H. Asanuma. 2007. Induced seismicity associated with enhanced geothermal systems. *Geothermics* 36:185-222.

Majer, E.L., J. Nelson, A. Robertson-Tait, J. Savy, and I. Wong. 2012. Protocol for Addressing Induced Seismicity Associated with Enhanced Geothermal Systems. DOE/EE-0662. U.S. Department of Energy. Available at www1.eere.energy.gov/geothermal/pdfs/geothermal_seismicity_protocol_012012.pdf (accessed April 2012).

Zoback, M.D. 2012 (April 2). Managing the seismic risk posed by wastewater disposal. *Earth Magazine* 38-43. Available at nodrilling.files.wordpress.com/2012/02/zoback-earth.pdf (accessed April 2012).

CHAPTER SEVEN

Addressing Induced Seismicity: Findings, Conclusions, Research, and Proposed Actions

Induced seismic activity attributed to a range of human activities has been documented since at least the 1920s. However, recent induced seismic events related to energy technology development projects that involve fluid injection or withdrawal in the United States have drawn heightened public attention. Although none of these events resulted in loss of life or significant damage, their effects were felt by local residents. These induced seismic events, though usually small in scale, can be disturbing for the public and raise concern about additional seismic activity and its consequences in areas where energy development is ongoing or planned. The findings, gaps, proposed actions, and research recommendations outlined in this chapter, based upon material presented earlier in the report, address

- the types and causes of induced seismicity;
- issues specific to each energy technology addressed in the study (geothermal energy, conventional and unconventional oil and gas production, injection wells for disposal of wastewater associated with energy development, and carbon capture and storage [CCS]);
- oversight, monitoring, and coordination of underground injection activities to help avoid felt induced seismicity;
- hazards and risk assessment; and
- best practices.

Although credible and viable research into possible induced seismic events has been conducted to date by industry, the academic community, and the federal government, further research is required because of the potential controversies surrounding such events. The Department of Energy, the U.S. Geological Survey, and the National Science Foundation are important organizations both for conducting and for supporting this kind of research and research partnerships with industry and academia. In addition to proposed actions to address induced seismicity, research recommendations are specifically highlighted in Box 7.1; some of these recommendations are specific to individual energy technologies, but most can be conducted with a purpose to understand induced seismicity more broadly.

BOX 7.1
Research Recommendations

Data Collection—Field and Laboratory

1. Collect, categorize, and evaluate data on potential induced seismic events in the field. High-quality seismic data are central to this effort. Research should identify the key types of data to be collected and data collection protocol.
2. Conduct research to establish the means of making in situ stress measurements nondestructively.
3. Conduct additional field research on microseisms in natural fracture systems including field-scale observations of the very small events and their native fractures.
4. Conduct focused research on the effect of temperature variations on stressed jointed rock systems. Although of immediate relevance to geothermal energy projects, the results would benefit understanding of induced seismicity in other energy technologies.
5. Conduct research that might clarify the in situ links among injection rate, pressure, and event size.

Instrumentation

1. Conduct research to address the gaps in current knowledge and availability of instrumentation: Such research would allow the geothermal industry, for example, to develop this domestic renewable source more effectively for electricity generation.

Hazard and Risk Assessment

1. Direct research to develop steps for hazard and risk assessment for single energy development projects (as described in Chapter 5, Table 5.2).

TYPES AND CAUSES OF INDUCED SEISMICITY

Findings

1. The basic mechanisms that can induce seismicity related to energy-related injection and extraction activities are not mysterious and are presently well understood.
2. Only a very small fraction of injection and extraction activities among the hundreds of thousands of energy development wells in the United States have induced seismicity at levels that are noticeable to the public.
3. Current models employed to understand the predictability of the size and location of earthquakes through time in response to net fluid injection or withdrawal

Modeling

1. Identify ways in which simulation models can be scaled appropriately to make the required predictions of the field observations reported.
2. Conduct focused research to advance development of linked geomechanical and earthquake simulation models that could be utilized to better understand potential induced seismicity and relate this to number and size of seismic events.
3. Use currently available and new geomechanical and earthquake simulation models to identify the most critical geological characteristics, fluid injection or withdrawal parameters, and rock and fault properties controlling induced seismicity.
4. Develop simulation capabilities that integrate existing reservoir modeling capabilities with earthquake simulation modeling for hazard and risk assessment. These models can be refined on a probabilistic basis as more data and observations are gathered and analyzed.
5. Continue to develop capabilities with coupled reservoir fluid flow and geomechanical simulation codes to understand the processes underlying the occurrence of seismicity after geothermal wells have been shut in; the results may also contribute to understanding post-shut-in seismicity in relation to other energy technologies.

Research Specific to CCS with Potential to Understand Induced Seismicity Broadly

1. Use some of the many active fields where CO_2 flooding for enhanced oil recovery (EOR) is conducted to understand more about the apparent lack of felt induced seismic events in these fields; because CO_2 is compressible in the gaseous phase are other factors beyond pore pressure important to understand in terms of CCS?
2. Develop models to estimate the potential earthquake magnitude that could be induced by large-scale CCS.
3. Develop detailed physicochemical and fluid mechanical models for injection of supercritical CO_2 into potential storage aquifers.

require calibration from data from field observations. The success of these models is compromised in large part due to the lack of basic data at most locations on the interactions among rock, faults, and fluid as a complex system.

4. Increase of pore pressure above ambient value due to injection of fluids or decrease in pore pressure below ambient value due to extraction of fluids has the potential to produce seismic events. For such activities to cause these events, a certain combination of conditions has to exist simultaneously:
   a. Significant change in net pore pressure in a reservoir
   b. A preexisting, near-critical state of stress along a fracture or fault that is determined by crustal stresses and the fracture or fault orientation
   c. Fault-rock properties supportive of brittle failure

5. Independent capability exists for geomechanical modeling of pore pressure, temperature, and rock stress changes induced by injection and extraction and for modeling of earthquake sequences given knowledge of stress changes, pore pressure changes, and fault characteristics.
6. The range of scales over which significant responses arise in the Earth with respect to induced seismic events is very wide and challenges the ability of models to simulate and eventually predict observations from the field.

Gaps

1. The basic data on fault locations and properties, in situ stresses, pore pressures, and rock properties are insufficient to implement existing models with accuracy on a site-specific basis.
2. Current predictive models cannot properly quantify or estimate the seismic efficiency and mode of failure; geomechanical deformation can be modeled, but a challenge exists to relate this to number and size of seismic events.

Proposed Actions

The actions proposed to advance understanding of the types and causes of induced seismicity involve research recommendations outlined in Box 7.1. These recommendations also have relevance for specific energy technologies and address gaps in understanding induced seismicity.

ENERGY TECHNOLOGIES: HOW THEY WORK

Overarching Findings for All Technologies

1. Injection pressures and net fluid volumes in energy technologies, such as geothermal energy and oil and gas production, are generally controlled to avoid increasing pore pressure in the reservoir above the initial reservoir pore pressure. These technologies thus appear less problematic in terms of inducing felt seismic events than technologies that result in a significant net increase or decrease in fluid volume.
2. The basic data needed to fully evaluate the potential for induced seismicity—including fault locations and properties, in situ stresses, fluid pressures, and rock properties—are very difficult and expensive to obtain.
3. Existing regional seismic arrays may not be capable of precisely locating small induced seismic events to determine causality and better establish the characteristics of induced seismicity.

4. Temporary local seismic arrays can be installed to find faults, determine source mechanisms, decrease error in location of seismic events, and increase resolution of future events.

GAP

Simple geometric considerations to help visualize subsurface problems and identify cases that deserve further attention are in most cases absent. Developing these kinds of simple analyses could, for example, be applied to understand the length scale affected by a single well or by multiple wells relative to depth or proximity to major faults and to the surface.

PROPOSED ACTION

In locales where a causal relationship may exist between subsurface energy activities and seismicity (even for small earthquakes of **M** between 3 and 4), a local seismic array should be installed for seismic monitoring. An appropriate body to determine whether such an array is necessary may be the permitting agency for the well(s) thought to be involved in the seismicity. Installation of such an array may require significant resources (including instrumentation and analysis). Existing groups, such as the U.S. Geological Survey, national laboratories, state geological surveys, universities, and private companies have the expertise necessary to install arrays and conduct the necessary analyses. Full disclosure of the data and results of such monitoring is required.

Geothermal Energy

FINDINGS

1. The induced seismic responses to injection differ in cause and magnitude with each of the three different forms of geothermal resources. At the vapor-dominated Geysers field hundreds of earthquakes of **M** 2 or greater are produced annually with one or two of **M** 4, all apparently caused principally by cooling and contraction of the reservoir rocks. The liquid-dominated field developments generally cause little if any induced seismicity because the water injection typically replaces similar quantities of fluid extracted at similar pressures and temperatures. The high-pressure hydraulic fracturing into generally impermeable rock associated with the stimulation operations at enhanced geothermal systems (EGS) projects can cause hundreds of small microseismic events and an occasional earthquake of up to **M** 3 due mainly to the imposed increased fluid pressures.

2. The mitigation of the effects of induced seismicity is in some instances clearly necessary to maintain or to restore public acceptance of the geothermal power generation activities. The early use of a "best practices" protocol and a "traffic light" control system indicates that such measures can provide an effective means to control operations so that the intensity of the induced seismicity is within acceptable levels. Further information on implementation of a protocol and control system is outlined under the final section in this chapter, Best Practices.

Gaps

1. Suitable coupled reservoir fluid flow and geomechanical simulation codes are not currently available to understand the processes underlying the occurrence of seismicity after geothermal wells have been shut in (ceased operation).
2. Field operators currently do not have ready access to downhole temperature and pressure recording instruments capable of making accurate measurements where reservoir conditions reach 750°F.

Proposed Actions

1. Adopt and use a matrix-style "best practices" protocol by developers as outlined in Chapter 6: Such a protocol is appropriate to use in those cases where there is a known probability of inducing seismicity at levels that could pose a concern to the public. In those cases where induced seismicity occurs but was previously unanticipated, the developer should consider adopting the protocol procedures needed to complete the project in a manner more satisfactory to the public.
2. Fully disclose and discuss a "traffic light" system in a public forum prior to the start of operations when such a system is to be adopted or imposed. Such disclosure and discussion will ensure that these safeguards are clearly known and understood by all concerned.

Conventional Oil and Gas Development Including Oil and Gas Withdrawal, Secondary Recovery, and Enhanced Oil Recovery

Findings

1. Generally, withdrawal associated with conventional oil and gas recovery has not caused significant seismic events; however, several major earthquakes have been associated with conventional oil and gas withdrawal.

2. Relative to the large number of waterflood projects for secondary recovery, the small number of documented instances of felt induced seismicity suggests such projects pose relatively small risk for events that would be of concern to the public.
3. The committee has not identified any documented, felt induced seismic events associated with EOR (tertiary recovery). The potential for induced seismicity is low in EOR operations as pore pressure is not significantly increased beyond the original levels in the reservoir because injected fluid volumes tend to be balanced by fluid withdrawals.

Unconventional Oil and Gas: Hydraulic Fracturing for Shale Gas Development

FINDINGS

1. The process of hydraulic fracturing a well as presently implemented for shale gas recovery does not pose a high risk for inducing felt seismic events. Thirty-five thousand wells have been hydraulically fractured for shale gas development to date in the United States. To date, hydraulic fracturing for shale gas production was cited as the possible cause of one case of felt seismic events in Oklahoma in 2011, the largest of which was **M** 2.8. The quality of the event locations was not adequate to fully establish a direct causal link to the hydraulic fracture treatment. Hydraulic fracturing for shale gas development has been confirmed as the cause of induced seismic events in one case worldwide—in Blackpool, England (maximum **M** 2.3).
2. One case of induced seismicity (maximum **M** 1.9) was documented in Oklahoma in the late 1970s as being caused by hydraulic fracturing for oil and gas development for conventional oil and gas extraction.

PROPOSED ACTION

When a seismic event occurs that appears to be associated with hydraulic fracturing and is considered to be a concern to the health, safety, and welfare of the public, an assessment is needed to understand the causes of the seismicity (see protocol that follows).

Injection Wells for the Disposal of Water Associated with Energy Extraction

FINDINGS

1. The United States currently has approximately 30,000 Class II wastewater disposal wells; very few felt induced seismic events have been reported as either caused by

or likely related to these wells. Rare cases of wastewater injection have produced seismic events, typically less than **M** 5.0.

2. Injected fluid volume, injection rate, injection pressure, and proximity to existing faults and fractures are factors that determine the probability to create a seismic event. High injection volumes in the absence of corresponding extractions may increase pore pressure and in proximity to existing faults could lead to an induced seismic event.
3. The area of potential influence from injection wells may extend over several square miles, and induced seismicity may continue for months to years after injection ceases.
4. Reducing the injection volumes, rates, and pressures has been successful in decreasing rates of felt seismicity in cases where events have been induced.
5. Evaluating the potential for induced seismicity in the location and design of injection wells is difficult because no cost-effective way to locate unmapped faults and measure in situ stress currently exists.

Gaps

1. Effective and economical tools are not available to accurately predict induced seismic activity prior to injection.
2. No capability exists to predict exactly how reducing volumes, pressures, and rates can lead to reduction in seismicity after it has begun. The models discussed in Chapter 2 are critical to developing the capacity to make such predictions.

Proposed Actions

The actions proposed by the committee to address the potential for induced seismicity related to injection wells for disposal of wastewater are similar to those suggested for geothermal energy technologies:

1. The adoption and use of a matrix-style "best practices" protocol as outlined in Chapter 6 in those cases where there is a known probability of inducing seismicity at levels that could pose a concern to the public. In those cases where the need becomes apparent only after disposal has begun, the developer should adopt the protocol procedures needed to complete the project in a manner that protects public safety.
2. When a "traffic light" system is to be adopted or imposed to control operations that could cause unacceptable levels of induced seismicity, full disclosure and discussion of the system at a public forum is necessary prior to the start of opera-

tions. Knowledge and understanding of these safeguards by all concerned are of great importance. Further information is outlined under the final section in this chapter, Best Practices.

Carbon Capture and Storage

FINDINGS

1. The only long-term (~14 years) commercial CO_2 sequestration project in the world at the Sleipner field off the shore of Norway is of a small scale relative to commercial projects proposed in the United States. Extensive seismic monitoring at this offshore site has not indicated any significant induced seismicity.
2. Proposed injection volumes of liquid CO_2 in large-scale sequestration projects (> 1 million metric tonnes per year) are much larger than those associated with the other energy technologies currently being considered. There is no experience with fluid injection at these large scales and little data on seismicity associated with CO_2 pilot projects. If the reservoirs behave in a similar manner to oil and gas fields, these large volumes have the potential to increase the pore pressure over vast areas. Relative to other technologies, such large affected areas may have the potential to increase both the number and the magnitude of seismic events.
3. CO_2 has the potential to react with the host/adjacent rock and cause mineral precipitation or dissolution. The effects of these reactions on potential seismic events are not understood.

GAPS

1. The short- and long-term effects of supercritical CO_2 in influencing rock strength and rock slip strength are not well understood.
2. The potential earthquake magnitudes that can be induced by the injection volumes being proposed for CCS are not known.
3. The complexities of hydrochemical-mechanical effects on CO_2 injection and storage are not thoroughly understood.

PROPOSED ACTIONS

Because of the lack of experience with large-scale fluid injection for CCS, continued research supported by the federal government is needed on the potential for induced seismicity in large-scale CCS projects. Some specific research recommendations are outlined in Box 7.1. As part of a continued research effort, collaboration between federal agencies and foreign operators of CCS sites is important to understand induced seismic events and their effects on CCS operations.

OVERSIGHT, MONITORING, AND COORDINATION OF UNDERGROUND INJECTION ACTIVITIES FOR MITIGATING INDUCED SEISMICITY

Findings

1. Induced seismicity may be produced by a number of different energy technologies and may result from either injection or extraction of fluid. As such, responsibility for oversight of activities that can cause induced seismicity is dispersed among a number of federal and state agencies.
2. Recent, potentially induced seismic events in the United States have been addressed in a variety of manners involving local, state, and federal agencies, and research institutions. These agencies and research institutions may not have resources to address these unexpected events, and more events could stress this ad hoc system.
3. Currently the Environmental Protection Agency (EPA) has primary regulatory responsibility for fluid injection under the Safe Drinking Water Act; however, this act does not explicitly address induced seismicity. EPA appears to be addressing the issue of induced seismicity through a current study in consultation with other federal and state agencies.
4. The U.S. Geological Survey (USGS) has the capability and expertise to address monitoring and research associated with induced seismic events. However, the scope of its mission within the seismic hazard assessment program is focused on large-impact, natural earthquakes. Significant new resources would be required if the USGS mission is expanded to include comprehensive monitoring and research on induced seismicity.

Gap

Mechanisms are lacking for efficient coordination of governmental agency response to seismic events that may have been induced.

Proposed Actions

1. In order to move beyond the current ad hoc approach for responding to induced seismicity, relevant agencies including EPA, USGS, land management agencies, and possibly the Department of Energy, as well as state agencies with authority and relevant expertise (e.g., oil and gas commissions, state geological surveys, state

environmental agencies, etc.) should consider developing coordination mechanisms to address induced seismic events that correlate to established best practices (see recommendation below).

2. Appropriating authorities and agencies with potential responsibility for induced seismicity should consider resource allocations for responding to induced seismic events in the future.

HAZARDS AND RISK ASSESSMENT

Gap

Currently, methods do not exist to implement assessments of hazards upon which risk assessments depend. The types of information and data required to provide a robust hazard assessment would include

- net pore pressures, in situ stresses, and information on faults;
- background seismicity; and
- gross statistics of induced seismicity and fluid injection for the proposed site activity.

Proposed Actions

1. A detailed methodology should be developed for quantitative, probabilistic hazard assessments of induced seismicity risk. The goals in developing the methodology would be to
    - make assessments before operations begin in areas with a known history of felt seismicity and
    - update assessments in response to observed induced seismicity.
2. Data related to fluid injection (well location coordinates, injection depths, injection volumes and pressures, time frames) should be collected by state and federal regulatory authorities in a common format and made accessible to the public (through a coordinating body such as the USGS).
3. In areas of high density of structures and population, regulatory agencies should consider requiring that data to facilitate fault identification for hazard and risk analysis be collected and analyzed before energy operations are initiated.

BEST PRACTICES

Findings

1. The DOE Protocol for EGS, which lists seven sequential steps, provides a reasonable initial model for dealing with induced seismicity that can serve as a template for other energy technologies.
2. Based on this initial model, the committee has proposed two matrix-style protocols as examples to illustrate the manner in which these seven activities can ideally be undertaken concurrently (rather than only sequentially), while also illustrating how these activities should be adjusted as a project progresses from early planning through operations to completion.

Gap

No best practices protocol for addressing induced seismicity is generally in place for each of these technologies, with the exception of the protocol recently developed for EGS. The committee suggests that best practices protocols be adapted and tailored to each technology to allow continued energy technology development. Actions toward developing these protocols are outlined below.

Proposed Actions

1. A matrix-style "best practices" protocol should be developed in coordination with the permitting agency or agencies by experts in the field of each energy technology, including EOR, shale gas production, and CCS.
2. The adoption and use of such protocols by developers are recommended in each case where there is a known or substantial probability of inducing seismicity at levels that could pose a concern to the public. In cases where induced seismicity becomes an issue at some stage in the project, the developer can adopt the protocol procedures needed to continue the project in a manner more satisfactory to the public.
3. Even with the adoption and use of a best practices protocol, induced seismicity of serious concern to public health and safety may occur. The regulatory body affiliated with the permitting of well(s) should include, as part of each project's operation permit, a mechanism (such as a "traffic light" mechanism) for the well operator to be able to control, reduce, or eliminate the potential for felt seismic events.
4. When a traffic light system is to be adopted or imposed to control operations that may cause unacceptable levels of induced seismicity, full disclosure and discussion

of the adopted system at a public forum prior to the start of operations is advised so that these safeguards are clearly known and understood by all concerned. Simultaneous development of public awareness programs by federal or state agencies in cooperation with industry and the research community could aid the public and local officials in understanding and addressing the risks associated with small-magnitude induced seismic events.

Appendixes

Committee and Staff Biographies

COMMITTEE BIOGRAPHIES

Murray W. Hitzman (Chair) has been with Colorado School of Mines since 1996 as the Fogarty Professor of Economic Geology. In 2002 he was named Head of the Department of Geology and Geological Engineering. Prior to coming to academia he spent 11 years in the minerals industry. In addition to discovering the carbonate-hosted Lisheen Zn-Pb-Ag deposit in Ireland, he worked on porphyry copper and other intrusive-related deposits, precious metal systems, volcanogenic massive sulfide deposits, sediment-hosted Zn-Pb and Cu deposits, and iron oxide Cu-U-Au-LREE deposits throughout the world. He spent 2½ years in Washington, D.C., working first in the U.S. Senate and later in the White House Office of Science and Technology Policy on environmental and natural resource issues. He has received numerous awards and has published approximately 100 papers. His current interest focuses on deposit- and district-scale studies of metallic ore systems and on social license issues in mining. Dr. Hitzman was a member of the National Research Council's Panel on Technologies for the Mining Industries, and he was a member of Committee on Earth Resources for two 3-year terms prior to becoming chair for a 3-year term in 2004. He received his Ph.D. in geology from Stanford University in 1983.

Donald D. Clarke has worked for the past 6 years as a geological consultant for a variety of private firms and city governments in Southern California, focusing on geological evaluations of oil fields. Part of his current portfolio also includes a CO_2 sequestration project. Prior to establishing his consultancy, he worked for more than 2 decades with the Department of Oil Properties of the City of Long Beach, California, retiring from his position as Division Engineer and Chief Geologist in 2004. During his time with the City of Long Beach, he worked extensively on the giant Wilmington oil field and the California offshore. Mr. Clarke began his career in 1974 as an energy and mineral resources engineer with the California State Lands Commission. His strong interests in community outreach and education have been demonstrated over the years through his teaching geology at Compton Community College, serving on the board of directors for the Petroleum Technology Transfer Council, and serving on and chairing numerous advisory councils and committees of the American Association of Petroleum Geologists (AAPG). A member of AAPG since 1986, he served as Pacific Section AAPG President, was elected to be Chairman of the AAPG House of Delegates, and has received numerous AAPG awards, including the Distinguished Service Award in 2002. He also served on the National Research Council committee that produced the 2002 report *Geoscience Data and Collections: National Resources in Peril.* In the last year he appeared and served as an advisor for the Swiss movie, *A Crude Awakening*; the National Geographic show, *Gallon of Gas* (part of the Man Made Series); and the VBS TV show *LA's Hidden Wells.* This past summer he was interviewed by the Canadian Broadcasting Corporation and Spiegel Television (Germany) about oil development in the Los Angeles area. Mr. Clarke has published or presented more than 50 technical papers on topics that include computer mapping, sequence stratigraphy, horizontal drilling, structural geology, and reservoir evaluation, and he has been recognized by the Institute for the Advancement of Engineering as a fellow. He received his B.S. in geology from California State University, Northridge, with additional graduate study at California State University, Northridge, Los Angeles, and Long Beach.

Emmanuel Detournay is a professor of geomechanics in the Department of Civil Engineering at the University of Minnesota. He also holds a joint appointment with Commonwealth Scientific and Industrial Research Organisation Earth Science and Resource Engineering, where he leads the Drilling Mechanics Group. Prior to his current positions, he was senior research scientist at Schlumberger Cambridge Research in England. His expertise is in petroleum geomechanics with two current areas of focus: mechanics of hydraulic fractures and drilling mechanics. He has authored about 160 papers. He also has been awarded six U.S. patents and has received several scientific awards for his work. Dr. Detournay received his M.S. and Ph.D. in geoengineering from the University of Minnesota.

James H. Dieterich (NAS) is a distinguished professor of geophysics at the University of California, Riverside. His research has led to a new understanding of the Earth's crust. He is an internationally renowned authority in rock mechanics, seismology, and volcanology. His pioneering studies in the theory, measurement, and application of frictional processes in rocks have had major implications for predicting fault instability and earthquake nucleation. His previous work on the rate- and state-dependent representation of fault constitutive properties is now being applied in modeling of seismicity, including aftershocks and triggering of earthquakes, and in inverse models that use earthquake rates to map stress changes in space and time. Dr. Dieterich recently launched a new effort to investigate fault slip and earthquake processes in geometrically complex fault systems, which includes development of large-scale quasidynamic simulations of seismicity in fault systems, and investigation of the physical interactions and stressing conditions that control system-level phenomena. Dr. Dieterich received his Ph.D. in geology and geophysics from Yale University.

David K. Dillon is the principal of David K. Dillon PE, LLC, a petroleum engineering consulting firm located in Centennial, Colorado. He holds a B.S. degree in civil engineering from the University of Colorado at Boulder (1974). He is a licensed professional engineer in Colorado (#19171) and Wyoming (#12530) and has been a member of the Society of Petroleum Engineers for over 35 years. Before starting his career as a consulting engineer, Mr. Dillon worked in the private oil and gas industry for 20 years as a drilling engineer, a production engineer, and a reservoir engineer. He has extensive experience in optimizing production from existing oil and gas fields, secondary recovery operations, and the calculation of oil and gas reserves. Mr. Dillon was also an Engineering Supervisor and the Engineering Manager for the Colorado Oil and Gas Conservation Commission for over 15 years. The Colorado Oil and Gas Conservation Commission is the regulating body for oil and gas drilling and production in the state of Colorado. As the Engineering Manager he was instrumental in the drafting and adoption of new rules by the Commission and the review and approval of underground injection permits for the State of Colorado. Mr. Dillon has offered expert testimony before the oil and gas commissions of several states.

Sidney J. Green (NAE) is research professor at the University of Utah, where he holds a dual appointment in mechanical engineering and civil and environmental engineering. He is also a Schlumberger Senior Advisor and was one of the founders and former President and Chief Executive Officer of TerraTek, a geomechanics engineering firm, which was acquired by Schlumberger in 2006. Mr. Green has worked in the area of geomechanics for nearly 5 decades. He has published numerous papers and reports, holds a number of patents, has given many presentations on geomechanics, and has received a number of rock mechanics and geomechanics recognitions. He has served on government committees and on many university and national laboratory advisory boards, and he has testified at a number of congressional hearings. He has served as member of the board of directors for a number of businesses. He received the Outstanding Engineer award and the Entrepreneur of the Year award from Utah, and the Distinguished Alumni Award (1976) and the Professional Degree recognition (1998) from the former Missouri School of Mines. He received the 1989 Honorary Alumni Award and the 2009 Engineering Achievement Award from the University of Utah. He is a past member of the Greater Salt Lake Chamber of Commerce Board of Governors and was recently elected a Fellow of the American Rock Mechanics Association. He is a member of the U.S. National Academy of Engineering. He most recently served as a member of the NRC Committee on Assessment of the Department of Energy's Methane Hydrate Research and Development Program: Evaluating Methane Hydrate as a Future Energy Resource. Mr. Green has a B.S. from the former Missouri School of Mines and an M.S. from the University of Pittsburgh, both in mechanical engineering. He attended 1 year at Pennsylvania State University graduate school and 2 years at Stanford University, where he received the degree of engineer in engineering mechanics.

Robert M. Habiger worked for ConocoPhillips for over 28 years in various scientific and management capacities in the disciplines of petrophysics and geophysics. While there, he held various positions in research and development and in international exploration, including Manager for Seismic Technology in the Houston corporate offices. He joined Spectraseis as Chief Technology Officer in February 2007, where he is responsible for all technical aspects of the company's research and commercial offerings in passive seismic technology. These programs and products include both hydrocarbon reservoir fluids monitoring from low-frequency passive seismic and microseismic monitoring associated with hydraulic fracturing and fluid injection/removal. Rob is the Director of the Low Frequency Seismic Partnership, an industrial research consortium studying the application of low-frequency passive seismic methods to hydrocarbon fluid mapping. He holds bachelor's, master's, and Ph.D. degrees in physics.

Robin K. McGuire (NAE) is a consulting engineer specializing in earthquake engineering, risk analysis, and decision analysis. His experience includes directing projects to determine earthquake design requirements for new nuclear power plants in the central and eastern United States; making recommendations to the Electric Power Research Institute and the U.S. Nuclear Regulatory Commission on seismic design requirements; consulting for the National Committee on Property Insurance on earthquake matters and making recommendations to the California Department of Insurance; serving as lead consultant on probabilistic performance assessment of the Yucca Mountain site as a possible high-level waste repository; and consulting on numerous U.S. and overseas studies of seismic and environmental risk for utilities, insurance groups, and commercial clients. Dr. McGuire was president of the Seismological Society of America (SSA) in 1991-1992, authored the book *Seismic Hazard and Risk Analysis* in 2004, and was the Joyner Lecturer in 2009 for the Earthquake Engineering Research Institute and the SSA. Dr. McGuire received his S.B. in civil engineering from the Massachusetts Institute of Technology, his M.S. in structural engineering from the University of California, Berkeley, and his Ph.D. in structural engineering from the Massachusetts Institute of Technology.

James K. Mitchell (NAS/NAE) is currently University Distinguished Professor Emeritus at Virginia Polytechnic Institute and State University (Virginia Tech) and Consulting Geotechnical Engineer. Prior to joining Virginia Tech in 1994, he served on the faculty at the University of California, Berkeley, since 1958, holding the Edward G. Cahill and John R. Cahill Chair in the Department of Civil and Environmental Engineering there at the time of his retirement in 1993. Concurrent to his tenure at UC Berkeley, he was Chairman of Civil Engineering from 1979 to 1984. His primary research activities have focused on experimental and analytical studies of soil behavior related to geotechnical problems, admixture stabilization of soils, soil improvement and ground reinforcement, physicochemical phenomena in soils, environmental geotechnics, time-dependent behavior of soils, in situ measurement of soil properties, and mitigation of ground failure risk during earthquakes. He has authored more than 375 publications, including the graduate-level text and geotechnical reference *Fundamentals of Soil Behavior*. A licensed civil engineer and geotechnical engineer in California and professional engineer in Virginia, Dr. Mitchell has served as chairman or officer for numerous national and international organizations. He has chaired the NRC Geotechnical Board and three NRC study committees, and served as a member of several other NRC study committees. He has received numerous awards, including the Norman Medal and the Outstanding Projects and Leaders Award from the American Society of Civil Engineers, and the NASA Medal for Exceptional Scientific Achievement. He was elected to the National Academy of Engineering in 1976 and to the National Academy of Sciences in 1998. Dr. Mitchell received a bachelor of civil engineering from Rensselaer Polytechnic Institute, and M.S. and doctor of science degrees in civil engineering from the Massachusetts Institute of Technology.

Julie E. Shemeta is the president and founder of MEQ Geo Inc., a microseismic consulting and services company based in Denver, Colorado. She has worked on microseismic projects in North America, Australia, and India, including hydraulic fracture monitoring in tight gas, shale gas and oil, steam-assisted gravity drainage, and coalbed methane projects. Her background includes deep-water oil and gas exploration in the Gulf of Mexico, working in the geothermal industry for developments in Indonesia and the Philippines, and working for a microseismic vendor providing data processing and consulting on hydraulic fracture monitoring. Ms. Shemeta has been actively involved with the development of software for both processing and visualization of microseismic throughout her 20-year career. She has served on numerous meeting committees for the Society of Exploration Geophysicists, the Society of Petroleum Engineers, and the AAPG. She co-chaired the DGS/RMAG (Denver Geophysical Society and Rocky Mountain Association of Geologists) 3-D Seismic Symposium from 2009 to 2011 and is still active on the committee. She served as the Denver Geophysical Society Treasurer in 2008-2009. She obtained her B.S. in geology at the University of Washington and her M.S. in geophysics with a specialty in earthquake seismology at the University of Utah.

John L. (Bill) Smith is presently a geothermal consultant having recently retired as a senior geologist at the Northern California Power Agency (NCPA). He has 46 years of diversified geologic, geophysical, and geochemical experience in the geothermal and oil and gas industry, including numerous geothermal exploration and development projects in the western United States and Japan. For the past 25 years he has worked at The Geysers, first designing, permitting, and evaluating steam production and water injection wells to initially supply a 220 MW power project, and then for more than the past decade monitoring the induced seismicity that occurs both within the NCPA area of operations and throughout the entire Geysers field. Prior to joining The Geysers, Dr. Smith worked for 10 years as an oil and gas exploration geologist and geophysicist (seismologist) for Standard Oil of California (Chevron), then for 11 years as Vice President of Exploration for Republic Geothermal, which included geothermal exploration and development projects throughout California, Nevada, Utah, and Japan. Dr. Smith received his A.B. in geology from Middlebury College and his M.A. and Ph.D. in geological sciences from Indiana University.

STAFF BIOGRAPHIES

Elizabeth A. Eide is director of the Board on Earth Sciences and Resources at the NRC. Prior to joining the NRC as a staff officer in 2005, she served as a researcher, team leader, and laboratory manager for 12 years at the Geological Survey of Norway in Trondheim. In Norway her research included basic and applied projects related to isotope geochronology, mineralogy and petrology, and crustal processes. Her publications include more than 40 journal articles and book chapters, and 10 Geological Survey reports. She has overseen 10 NRC studies. She completed a Ph.D. in geology at Stanford University and received a B.A. in geology from Franklin and Marshall College.

Courtney Gibbs is a program associate with the NRC Board on Earth Sciences and Resources. She received her degree in graphic design from the Pittsburgh Technical Institute in 2000 and began working for the National Academies in 2004. Prior to her work with the board, Ms. Gibbs supported the Nuclear and Radiation Studies Board and the former Board on Radiation Effects Research.

Jason R. Ortego is a research associate with the Board on Earth Sciences and Resources at the National Academies. He received a B.A. in English from Louisiana State University in 2004 and an M.A. in international affairs from George Washington University in 2008. He began working for the National Academies in 2008 with the Board on Energy and Environmental Systems, and in 2009 he joined the Board on Earth Sciences and Resources.

Nicholas D. Rogers is a financial and research associate with the National Research Council Board on Earth Sciences and Resources. He received a B.A. in history, with a focus on the history of science and early American history, from Western Connecticut State University in 2004. He began working for the National Academies in 2006 and has primarily supported the board on a broad array of Earth resources, mapping, and geographical sciences issues.

Meeting Agendas

MEETING 1

Washington, DC, April 26-27, 2011

DAY ONE

08:00-09:00	CLOSED SESSION (Committee & NRC Staff only)
09:00-09:15	***Doors open; registration***
09:15-15:00	**OPEN SESSION—PUBLIC WELCOME TO ATTEND**
09:15-09:30	Welcome and introductions *Murray Hitzman, Chair*
09:30-15:00	**Presentations**
09:30-10:30	**Department of Energy** **George Guthrie**, Office of Fossil Energy/National Energy Technology Laboratory **JoAnn Milliken and Jay Nathwani**, Geothermal Technologies Program
10:30-11:00	**Allyson Anderson**, Professional staff, U.S. Senate Energy and Natural Resources Committee
11:00-11:15	Break
11:15-12:00	**Ernie Majer**, Senior Advisor to the ESD Director and Energy Program Leader, Lawrence Berkeley National Laboratory
12:00-13:00	*Lunch*
13:00-13:45	**Cliff Frohlich**, Professor, University of Texas at Austin

13:45-14:30 **Domenico Giardini**, Director, Swiss Seismological Service

14:30-15:00 General discussion *Murray Hitzman, Chair*

End of open session

15:00-17:00 CLOSED SESSION (Committee & NRC Staff only)

End of session

DAY TWO

08:00-13:30 CLOSED SESSION (Committee & NRC Staff only)

End of meeting

MEETING 2

The Geysers, CA, and Lawrence Berkeley National Laboratory, CA, July 13-15, 2011

DAY ONE

Committee members tour Geysers, led by representatives from NCPA and Calpine

DAY TWO

09:15-16:45 OPEN SESSION—PUBLIC WELCOME TO ATTEND

09:15-09:25 Welcome and introduction to study *Murray Hitzman, Chair*

09:25-12:30 Panel discussions

09:25-10:15 **Panel 1—Vapor-dominated geothermal resource development**
Melinda Wright, Calpine Corporation
Craig Hartline, Calpine Corporation
Bill Smith, Northern California Power Agency

10:15-10:45	**Panel 2—Liquid-dominated geothermal resource development** Charlene Wardlow, Ormat
10:45-11:00	*Break*
11:00-12:30	**Panel 3—EGS resource development** Mark Walters, Calpine Corporation Julio Garcia, Calpine Corporation Susan Petty, Chief Technology Officer, AltaRock Energy Inc. Ernst Huenges, Head of Reservoir Technologies, GFZ Potsdam Jay Nathwani, Department of Energy Geothermal Technologies Program
12:20-13:30	Lunch presentation— Ernie Majer, Lawrence Berkeley National Laboratory, on the topic of the Department of Energy Induced Seismicity Protocol
13:30-16:30	**Presentations**
13:30-14:00	**Federal land management** Linda Christian, Bureau of Land Management Oregon/Washington
14:00-15:00	**Community contributions** Mark Dellinger, Jeffrey Gospe, Hamilton Hess, Meriel Medrano, Cheryl Engels
15:00-15:15	*Break*
15:15-16:30	**Research** David Oppenheimer, USGS Jean Savy, Lawrence Berkeley National Laboratory
16:30-17:00	General discussion *Murray Hitzman, Chair*

End of open session

DAY THREE

08:00-12:00 CLOSED SESSION (Committee & NRC Staff only)

End of meeting

MEETING 3

Irvine, CA, August 18, 2011

08:30-14:15 OPEN SESSION—PUBLIC WELCOME TO ATTEND

08:30-08:40 Welcome and introduction to study *Murray Hitzman, Chair*

08:45-15:00 Presentations (presentations + time for discussion)

08:45-10:00 **Ola Eiken and Philip Ringrose,** *Statoil AS*
CO_2 sequestration and monitoring activities offshore Norway
Overview of CO_2 Monitoring Activities Offshore Norway (Sleipner, Snøhvit)—*Ola Eiken*
Future plans for microseismic and surface monitoring onshore and offshore—*Philip Ringrose*

10:00-10:15 Break

10:15-11:15 **James Rutledge,** *Los Alamos National Laboratory*

11:15-12:30 **Mark Zoback,** *Stanford University*
The potential for triggered seismicity associated with CO_2 sequestration and shale gas development

12:30-13:15 Lunch

13:15-14:15 **Michael Bruno,** *Terralog Technologies*

End of open session

MEETING 4

Dallas, TX, September 14-15, 2011

DAY ONE

07:30-08:15 CLOSED SESSION—COMMITTEE AND NRC STAFF ONLY

08:30-17:30 OPEN SESSION—PUBLIC WELCOME TO ATTEND

08:30-08:45 Welcome and Introductory Remarks *Murray Hitzman, Committee Chair*

Morning session moderated by Don Clarke and Jim Mitchell, Committee members

08:45-09:20 **Norm Warpinski**, Pinnacle—A Halliburton Service
Induced seismicity in shale stimulations

09:20-09:55 **Leo Eisner**, Czech Academy of Sciences and Seismik, Ltd.
Case examples of induced seismic events near shale gas operations

09:55-10:35 **Scott Ausbrooks**, Arkansas Geological Survey
Steve Horton, University of Memphis
Earthquakes in central Arkansas triggered by fluid injection at Class 2 UIC wells

10:35-10:50 Break

10:50-11:20 **John Jeffers**, Southwestern Energy
Observations and perspectives on induced seismicity and microseismicity associated with shale gas development

11:20-11:55 **Serge Shapiro**, Free University of Berlin
Quantitative understanding of induced microseismicity for reservoir characterization and development

11:55-12:30 **Doug Johnson**, Texas Railroad Commission
Regulatory response to induced seismicity in Texas

12:30-13:15 Lunch

Afternoon session moderated by David Dillon and Robin McGuire, Committee members

13:15-13:45 **Lisa Block**, Bureau of Reclamation
Deep injection of brine and monitored induced seismicity in Paradox Valley

13:45-14:15 **Philip Dellinger**, Environmental Protection Agency
Summary of EPA's current work with induced seismicity issues

14:15-14:50 **Shawn Maxwell**, Schlumberger
Overview of hydraulic fracture mapping

14:50-15:00 Break

15:00-15:40 **Rob Finley**, Illinois State Geological Survey
Midwest Geological Sequestration Consortium—Overview of approaches to induced seismicity

15:40-16:15 **Steve Melzer**, Melzer Consulting
Tertiary production and CO_2 enhanced oil recovery including conceptual risk of injection, reservoir surveillance, and sequestration monitoring

16:15-16:45 Wrap-up discussion *Moderated by Murray Hitzman*

End of Open Session

DAY TWO

07:45-09:45 CLOSED SESSION, COMMITTEE AND STAFF ONLY

10:00-13:00 OPEN SESSION—PUBLIC WELCOME TO ATTEND

10:00-10:10 Introductory Remarks *Murray Hitzman, Committee chair*

10:10-12:00 Panel discussion *Moderated by Julie Shemeta, Committee member*

Werner Heigl, Apache Corporation
Jamie Rich, Devon Energy

12:00-13:00 *Lunch*

End of open session

13:00-17:00 CLOSED SESSION, COMMITTEE AND NRC STAFF ONLY

DAY THREE

07:30-12:00 CLOSED SESSION, COMMITTEE AND STAFF ONLY

End of meeting

MEETING 5

Washington, DC, November 10-11, 2011

DAY ONE

08:00-09:30 CLOSED SESSION—COMMITTEE AND STAFF ONLY

09:30-10:45 OPEN SESSION—PUBLIC WELCOME

09:30-09:40 Welcome and Introductory Remarks *Murray Hitzman, Committee Chair*

09:40-10:00 **Allyson Anderson**, Professional staff, U.S. Senate Energy and Natural Resources Committee

10:00-10:15 **Jay Braitsch**, Department of Energy—Fossil Energy

10:15-10:30 **Jay Nathwani,** Department of Energy—Geothermal Technologies Program

10:30-10:45 General discussion

10:45-11:00 *Break*

End Open Session

11:00-20:00 CLOSED SESSION—COMMITTEE AND STAFF ONLY

DAY TWO

07:45-13:00 CLOSED SESSION—COMMITTEE AND STAFF ONLY

End of meeting

MEETING 5

Denver, CO, January 10-11, 2012

CLOSED SESSIONS—COMMITTEE AND STAFF ONLY

Observations of Induced Seismicity

Site/City/State	Country	Max Magnitude	Technology Type (causing induced seismicity)	Reference
Akmaar	Netherlands	3.5	Oil and gas extraction	Giardini (2011)
Akosombo	Ghana	5.3	Surface water reservoir	Guha (2000)
Apollo Hendrick Field, Texas	USA	2	Secondary recovery	Doser et al. (1992)
Ashtabula, Ohio	USA	3.6	Wastewater injection	Armbruster et al. (1987)
Assen	Netherlands	2.8	Oil and gas extraction	Grasso (1992)
Aswan	Egypt	5.6	Surface water reservoir	Guha (2000)
Attica, New York	USA	5.2	Other	Nicholson and Wesson (1992)
Bad Urach	Germany	1.8	Geothermal	Evans et al. (2012)
Bajina Basta	Yugoslavia	4.8	Surface water reservoir	Guha (2000)
Barsa-Gelmes-Wishka Oilfield	Turkmenistan	6	Secondary recovery	Kouznetsov et al. (1994)
Basel	Switzerland	3.4	Geothermal	Giardini (2011)
Belchalow	Poland	4.6	Other	Giardini (2011)
Benmore	New Zealand	5	Surface water reservoir	Guha (2000)
Bergermeer Field	Netherlands	3.5	Oil and gas extraction	van Eck et al. (2006)
Berlin	El Salvador	4.4	Geothermal	Bommer et al. (2006)
Bhatsa	India	4.8	Surface water reservoir	Guha (2000)
Blackpool	UK	2.3	Hydraulic fracturing	de Pater and Baisch (2011)

Site/City/State	Country	Max Magnitude	Technology Type (causing induced seismicity)	Reference
Cajuru, Brazil	Brazil	4.7	Surface water reservoir	Guha (2000)
Camarillas, Spain	Spain	4.1	Surface water reservoir	Guha (2000)
Canelles, Spain	Spain	4.7	Surface water reservoir	Guha (2000)
Catoosa, Oklahoma[1]	USA	4.7	Oil and gas extraction	Nicholson and Wesson (1992)
Cesano	Italy	2	Geothermal	Evans et al. (2012)
Charvak	Uzbekistan	4	Surface water reservoir	Guha (2000)
Clark Hill	USA	4.3	Surface water reservoir	Guha (2000)
Cleburne, Texas	USA	2.8	Oil and gas extraction	Howe et al. (2010)
Cleveland, Ohio[2]	USA	3	Other	Nicholson and Wesson (1992)
Coalinga, California	USA	6.5	Oil and gas extraction	McGarr (1991)
Cogdell Canyon Reef, Texas	USA	4.6	Secondary recovery	Davis and Pennington (1989); Nicholson and Wesson (1990)
Cold Lake, Alberta	Canada	2	Secondary recovery	Nicholson and Wesson (1990)
Cooper Basin	Australia	3.7	Geothermal	Majer et al. (2007)
Coso, California	USA	2.6	Geothermal	Julian et al. (2007); Foulger et al. (2008)
Coyote Valley	USA	5.2	Surface water reservoir	Guha (2000)
Dale, New York	USA	1	Other	Nicholson and Wesson (1990)
Dallas Fort Worth, Texas	USA	3.3	Wastewater injection	Frohlich et al. (2010)
Dan	Denmark	4	Oil and gas extraction	Grasso (1992)
Danjiangkou	China	4.7	Surface water reservoir	Guha (2000)
Denver, Colorado[3]	USA	4.8	Wastewater injection	Hermann et al. (1981)

Site/City/State	Country	Max Magnitude	Technology Type (causing induced seismicity)	Reference
Desert Peak, Nevada	USA	0.74	Geothermal	Chabora et al. (2012)
Dhamni	India	3.8	Surface water reservoir	Guha (2000)
Dollarhide, Texas	USA	3.5	Secondary recovery	Nicholson and Wesson (1992)
Dora Roberts, Texas	USA	3	Secondary recovery	Nicholson and Wesson (1992)
East Durant, Oklahoma	USA	3.5	Oil and gas extraction	Nicholson and Wesson (1992)
East Texas, Texas	USA	4.3	Secondary recovery	Nicholson and Wesson (1992)
Ekofisk	Norway	3.4	Oil and gas extraction	Grasso (1992)
El Dorado, Arkansas	USA	3	Wastewater injection	Cox (1991)
El Reno, Oklahoma[4]	USA	5.2	Oil and gas extraction	Nicholson and Wesson (1992)
Eola field, Oklahoma	USA	2.8	Hydraulic fracturing	Holland (2011)
Eucumbene	Australia	5	Surface water reservoir	Guha (2000)
Fashing, Texas	USA	3.4	Oil and gas extraction	Pennington et al. (1986)
Fenton Hill, New Mexico	USA	1	Geothermal	Nicholson and Wesson (1992)
Fjallbacka	Sweden	–0.2	Geothermal	Evans et al. (2012)
Fort St. John, British Columbia	Canada	4.3	Secondary recovery	Horner et al. (1994)
Foziling	China	4.5	Surface water reservoir	Guha (2000)
Gazli	Uzbekistan	7.3	Oil and gas extraction	Adushkin et al. (2000)
Geysers, California	USA	4.6	Geothermal	Majer et al. (2007)
Gobles Field, Ontario	Canada	2.8	Secondary recovery	Nicholson and Wesson (1990)
Goose Creek, Texas	USA	unknown[5]	Oil and gas extraction	Nicholson and Wesson (1992)
Grandval	France	unknown[6]	Surface water reservoir	Guha (2000)

Site/City/State	Country	Max Magnitude	Technology Type (causing induced seismicity)	Reference
Groningen Field	Netherlands	3	Oil and gas extraction	van Eck et al. (2006)
Gross Schonebeck	Germany	−1.1	Geothermal	Evans et al. (2012)
Grozny	Caucasus (Russia)	3.2	Oil and gas extraction	Guha (2000)
Gudermes	Caucasus (Russia)	4.5	Oil and gas extraction	Smirnova (1968)
Guy and Greenbrier, Arkansas	USA	4.7	Wastewater injection	Horton (2012)
Harz	Germany	3.5	Other	Giardini (2011)
Hellisheidi	Iceland	2.4	Geothermal	Evans et al. (2012)
Hijiori	Japan	0.3	Geothermal	Kaieda et al. (2010)
Hoover	USA	5	Surface water reservoir	Guha (2000)
Horstberg	Germany	0	Geothermal	Evans et al. (2012)
Hsinfengchiang	China	6.1	Surface water reservoir	Guha (2000)
Hunt Field, Mississippi[7]	USA	3.6	Secondary recovery	Nicholson and Wesson (1992)
Idukki	India	3.5	Surface water reservoir	Guha (2000)
Imogene Field, Texas	USA	3.9	Oil and gas extraction	Pennington et al. (1986)
Inglewood Oil Field, California	USA	3.7	Secondary recovery	Nicholson and Wesson (1992)
Ingouri	Caucasus (Russia)	4.4	Surface water reservoir	Guha (2000)
Itizhitezhi	Zambia	4.2	Surface water reservoir	Guha (2000)
Kariba	Zambia	6.2	Surface water reservoir	Guha (2000)
Kastraki	Greece	4.6	Surface water reservoir	Guha (2000)
Kermit Field, Texas	USA	4	Secondary recovery	Nicholson and Wesson (1990)

Site/City/State	Country	Max Magnitude	Technology Type (causing induced seismicity)	Reference
Kerr	USA	4.9	Surface water reservoir	Guha (2000)
Kettleman North, California	USA	6.1	Oil and gas extraction	McGarr (1991)
Keystone I Field, Texas	USA	3.5	Secondary recovery	Nicholson and Wesson (1990)
Keystone II Field, Texas	USA	3.5	Secondary recovery	Nicholson and Wesson (1990)
Kinnersani	India	5.3	Surface water reservoir	Guha (2000)
Koyna	India	6.5	Surface water reservoir	Guha (2000)
Krafla	Iceland	2	Geothermal	Evans et al. (2012)
Kremasta	Greece	6.3	Surface water reservoir	Guha (2000)
German Continental Deep Drilling Program	Germany	1.2	Geothermal	Evans et al. (2012)
Kurobe	Japan	4.9	Surface water reservoir	Guha (2000)
Kuwait	Kuwait	4.7	Oil and gas extraction	Bou-Rabee (1994)
Lacq	France	4.2	Oil and gas extraction	Grasso and Wittlinger (1990)
Lake Charles, Louisiana[8]	USA	3.8	Oil and gas extraction	Nicholson and Wesson (1990)
Lambert Field, Texas	USA	3.4	Secondary recovery	Nicholson and Wesson (1992)
Landau	Germany	2.7	Geothermal	Evans et al. (2012)
Larderello-Travale	Italy	3	Geothermal	Evans et al. (2012)
Latera	Italy	2.9	Geothermal	Evans et al. (2012)
LGDD	Russia	4.2	Other	Giardini (2011)
Love County, Oklahoma[9]	USA	2.8	Secondary recovery	Nicholson and Wesson (1990)

Site/City/State	Country	Max Magnitude	Technology Type (causing induced seismicity)	Reference
Love County, Oklahoma	USA	1.9	Oil and gas extraction (hydraulic fracturing for conventional oil and gas development)	Nicholson and Wesson (1990)
Manicouagan	Canada	4.1	Surface water reservoir	Guha (2000)
Marathon	Greece	5.7	Surface water reservoir	Guha (2000)
Matsushiro	Japan	2.8	Wastewater injection	Ohtake (1974)
Mica, Canada	Canada	4.1	Surface water reservoir	Guha (2000)
Monahans, Texas	USA	3	Secondary recovery	Nicholson and Wesson (1992)
Monte Amiata	Italy	3.5	Geothermal	Evans et al. (2012)
Montebello, California	USA	5.9	Oil and gas extraction	Nicholson and Wesson (1992)
Montecillo, South Carolina	USA	2.8	Surface water reservoir	Guha (2000)
Monteynard	France	4.9	Surface water reservoir	Guha (2000)
Mutnovsky, Kamchatka	Russia	2	Geothermal	Kugaenko et al. (2005)
Northern Panhandle, Texas	USA	3.4	Secondary recovery	Nicholson and Wesson (1990)
Nurek	Tadjikstan	4.6	Surface water reservoir	Guha (2000)
Ogachi	Japan	2	Geothermal	Kaieda et al. (2010)
Petroleum field	Oman	2.1	Oil and gas extraction	Sze (2005)
Orcutt Field, California	USA	3.5	Oil and gas extraction	Nicholson and Wesson (1992)
Oroville, California	USA	5.7	Surface water reservoir	Guha (2000)
Paradise Valley, Colorado	USA	0.8	Wastewater injection	Nicholson and Wesson (1992)

Site/City/State	Country	Max Magnitude	Technology Type (causing induced seismicity)	Reference
Paradox Valley, Colorado	USA	4.3	Wastewater injection	Ake et al. (2005)
Perry, Ohio	USA	2.7	Wastewater injection	Nicholson and Wesson (1992)
Piastra	Italy	4.4	Surface water reservoir	Guha (2000)
Pieve de Cadore	Italy	4.3	Surface water reservoir	Guha (2000)
Porto Colombia	Brazil	5.1	Surface water reservoir	Guha (2000)
Rangely, Colorado	USA	3.1	Secondary recovery	Nicholson and Wesson (1990)
Renqiu oil field	China	4.5	Secondary recovery	Genmo et al. (1995)
Richland County, Illinois[10]	USA	4.9	Oil and gas extraction	Nicholson and Wesson (1992)
Rocky Mountain House, Alberta	Canada	3.4	Oil and gas extraction	Wetmiller (1986)
Romashkino, Tartarstan	Russia	4	Secondary recovery	Adushkin et al. (2000)
Rongchang, Chongqing	China	5.2	Oil and gas extraction	Lei et al. (2008)
Rosemanowes,	UK	2	Geothermal	Evans et al. (2012)
Roswinkel Field	Netherlands	3.4	Oil and gas extraction	van Eck et al. (2006)
Rotenburg	Germany	4.5	Oil and gas extraction	Giardini (2011)
Sefia Rud	Iran	4.7	Surface water reservoir	Guha (2000)
Shandong	China	2.4	Secondary recovery	Shouzhong et al. (1987)
Shenwo	China	4.8	Surface water reservoir	Guha (2000)
Sleepy Hollow, Nebraska	USA	2.9	Oil and gas extraction	Rothe and Lui (1983)
Snipe Lake	Canada	5.1	Secondary recovery	Nicholson and Wesson (1992)
Soultz	France	2.9	Geothermal	Evans et al. (2012)

Site/City/State	Country	Max Magnitude	Technology Type (causing induced seismicity)	Reference
South-central Texas	USA	4.3	Oil and gas extraction	Davis et al. (1995)
Southern Alabama	USA	4.9	Secondary recovery	Gomberg and Wolf (1999)
Sriramsagar	India	3.2	Surface water reservoir	Guha (2000)
Starogroznenskoe Oilfield	Russia	4.7	Oil and gas extraction	Kouznetsov et al. (1994)
Strachan, Alberta	Canada	3.4	Oil and gas extraction	Grasso (1992)
Southwest of Elsenbach	Germany	5.8	Other	Giardini (2011)
Tomahawk Field, New Mexico	USA	Unknown[11]	Wastewater injection	Nicholson and Wesson (1992)
Torre Alfina	Italy	3	Geothermal	Evans et al. (2012)
Unterhaching	Germany	2.4	Geothermal	Evans et al. (2012)
Upper Silesian	Poland	4.45	Other	Giardini (2011)
Vajont	Italy	3	Surface water reservoir	Guha (2000)
Valhall and Ekofisk Oilfields	Norway	Unknown[12]	Secondary recovery	Zoback and Zinke (2002)
Varragamba	Australia	5.4	Surface water reservoir	Guha (2000)
Vogtland	Germany		Wastewater injection	Baisch et al. (2002)
Vouglans	France	4.4	Surface water reservoir	Guha (2000)
War Wink Field, Texas	USA	2.9	Oil and gas extraction	Doser et al. (1992)
Ward-Estes Field, Texas	USA	3.5	Secondary recovery	Nicholson and Wesson (1992)
Ward-South Field, Texas	USA	3	Secondary recovery	Nicholson and Wesson (1992)
West Texas	USA	3.1	Oil and gas extraction	Keller et al. (1987)
Whittier Narrows, California	USA	5.9	Oil and gas extraction	McGarr (1991)

Site/City/State	Country	Max Magnitude	Technology Type (causing induced seismicity)	Reference
Wilmington Field, California	USA	3.3	Oil and gas extraction	Kovach (1974)

NOTE: "Other" refers to, e.g., coal and solution mining.

[1] Nicholson and Wesson (1990, 1992) were not able to confirm that the cause of the earthquake was oil and gas extraction; waterflooding and waste disposal were also active in the area at the time.

[2] Nicholson and Wesson (1990, 1992) were not able to confirm the accuracy of the maximum magnitude of this event, which occurred at the turn of the 20th century (1898-1907).

[3] For the Denver earthquakes of 1967-1968, Healy et al. (1968) reported magnitudes up to **M** 5.3 on an unspecified scale that were derived from local instruments.

[4] Nicholson and Wesson (1992) were not able to confirm conclusively that the earthquake was caused by oil and gas extraction.

[5] Nicholson and Wesson (1992) were not able to confirm conclusively that the earthquake was caused by oil extraction or the magnitudes of the events that occurred in the 1920s. Note that this location is not plotted in the figures (maps) in Chapter 1.

[6] Guha (2000) describes the earthquake using Modified Mercalli Intensity (V), but does not indicate moment magnitude.

[7] Nicholson and Wesson (1990, 1992) were not able to confirm conclusively that the event(s) were due to waterflooding for secondary recovery.

[8] Nicholson and Wesson (1990) were not able to confirm conclusively that the event(s) were due to oil and gas extraction activities.

[9] Nicholson and Wesson (1990) were not able to confirm the maximum magnitude of the events at this site.

[10] Nicholson and Wesson (1990, 1992) were not able to confirm conclusively that the event(s) were due to oil extraction.

[11] Nicholson and Wesson (1992) were not able to confirm the maximum magnitude of the events at this site.

[12] Zoback and Zinke (2002) did not provide a maximum magnitude, although the events recorded and analyzed are described as "microseismic" events.

REFERENCES

Adushkin, V.V., V.N. Rodionov, S. Turuntnev, and A.E. Yodin. 2000. Seismicity in the oil field. *Oilfield Review* Summer:2-17.

Ake, J., K. Mahrer, D. O'Connell, and L. Block. 2005. Deep-injection and closely monitored induced seismicity at Paradox Valley, Colorado. *Bulletin of the Seismological Society of America* 95(2):664-683.

Armbruster, J.G., L. Seeber, and K. Evans. 1987. The July 1987 Ashtabula earthquake (mb) 3.6 sequence in northeastern Ohio and a deep fluid injection well. Abstract. *Seismological Research Letters* 58(4):91.

Baisch, S., M. Bohnhoff, L. Ceranna, Y. Tu, and H.-P. Harjes. 2002. Probing the crust to 9-km depth: Fluid-injection experiments and induced seismicity at the KTB superdeep drilling hole, Germany. *Bulletin of the Seismological Society of America* 92(6):2369-2380.

Bommer, J.J., S. Oates, J.M. Cepeda, C. Lindholm, J. Bird, R. Torres, G. Marroquin, and J. Rivas. 2006. Control of hazard due to seismicity induced by a hot fractured rock geothermal project. *Engineering Geology* 83:287-306.

Bou-Rabee, F. 1994. Earthquake recurrence in Kuwait induced by oil and gas extraction. *Journal of Petroleum Geology* 17(4):473-480.

Chabora, E., E. Zemach, P. Spielman, P. Drakos, S. Hickman, S. Lutz, K. Boyle, A. Falconer, A. Robertson-Tait, N.C. Davatzes, P. Rose, E. Majer, and S. Jarpe. 2012. Hydraulic Stimulation of Well 27-15, Desert Peak Geothermal Field, Nevada, USA. *Proceedings of the Thirty-Seventh Workshop on Geothermal Reservoir Engineering,* Stanford University, CA, January 30-February 1.

Cox, R.T. 1991. Possible triggering of earthquakes by underground waste disposal in the El Dorado, Arkansas area. *Seismological Research Letters* 62(2):113-122.

Davis, S.D., and W.D. Pennington. 1989. Induced seismic deformation in the Cogdell oil field of West Texas. *Bulletin of the Seismological Society of America* 79(5):1477-1494.

Davis, S.D., P. Nyffenegger, and C. Frohlich. 1995. The 9 April 1993 earthquake in south-central Texas: Was it induced by fluid withdrawal? *Bulletin of the Seismological Society of America* 85(6):1888-1895.

de Pater, C.J, and S. Baisch. 2011. Geomechanical Study of Bowland Shale Seismicity, Synthesis Report. Cuadrilla Resources Ltd. Available at http://www.cuadrillaresources.com/wp-content/uploads/2012/02/Geomechanical-Study-of-Bowland-Shale-Seismicity_02-11-11.pdf (accessed July 12, 2012).

Doser, D.I., M.R. Baker, M. Luo, P. Marroquin, L. Ballesteros, J. Kingwell, H.L. Diaz, and G. Kaip. 1992. The not so simple relationship between seismicity and oil production in the Permian Basin, West Texas. *Pure and Applied Geophysics* 139(3/4):481-506.

Evans, K.F., A. Zappone, T. Kraft, N. Deichmann, and F. Moia. 2012. A survey of the induced seismic responses to fluid injection in geothermal and CO_2 reservoirs in Europe. *Geothermics* 41:30-54.

Foulger, G.R., B.R. Julian, and F.C. Monastero. 2008. Seismic monitoring of EGS tests at the Coso geothermal area, California, using accurate MEQ locations and full moment tensors. *Proceedings of the Thirty-Third Workshop on Geothermal Reservoir Engineering*, Stanford University, CA, January 28-30.

Frohlich, C., C. Hayward, B. Stump, and E. Potter. 2010. The Dallas-Fort Worth earthquake sequence: October 2008-May 2009. *Bulletin of the Seismological Society of America* 101(1):327-340.

Genmo, Z., C. Huaran, M. Shuqin, and Z. Deyuan. 1995. Research on earthquakes induced by water injection in China. *Pure and Applied Geophysics* 145(1):59-68.

Giardini, D. 2011. Induced Seismicity in Deep Heat Mining: Lessons from Switzerland and Europe. Presentation to the National Research Council Committee on Induced Seismicity Potential in Energy Production Technologies, Washington, DC, April 26.

Gomberg, J., and L. Wolf. 1999. Possible cause for an improbable earthquake: The 1997 Mw 4.9 southern Alabama earthquake and hydrocarbon recovery. *Geology* 27(4):367-370.

Grasso, J.-R. 1992. Mechanics of seismic instabilities induced by the recovery of hydrocarbons. *Pure and Applied Geophysics* 139(3/4):506-534.

Grasso, J.-R., and G. Wittlinger. 1990. Ten years of seismic monitoring over a gas field. *Bulletin of the Seismological Society of America* 80:2450-2473.

Guha, S.K. 2000. *Induced Earthquakes.* Dordrecht, the Netherlands: Kluwer Academic Publishers.

Healy, J.H., W.W. Rubey, D.T. Griggs, and C.B. Raleigh. 1968. The Denver earthquakes. *Science* 161:1301-1310.

Herrmann, R.B., S.-K. Park, and C.-Y. Wang. 1981. The Denver earthquakes of 1967-1968. *Bulletin of the Seismological Society of America* 71(3):731-745.

Holland, A. 2011. Examination of Possibly Induced Seismicity from Hydraulic Fracturing in the Eola Field, Garvin County, Oklahoma. Oklahoma Geological Survey Open-File Report OF1-2011. Available at www.ogs.ou.edu/pubsscanned/openfile/OF1_2011.pdf (accessed July 12, 2012).

Horner, R.B., J.E. Barclay, and J.M. MacRae. 1994. Earthquakes and hydrocarbon production in the Fort St. John area of northeastern British Columbia. *Canadian Journal of Exploration Geophysics* 30(1):39-50.

Horton, S. 2012. Disposal of hydrofracking-waste fluid by injection into subsurface aquifers triggers earthquake swarm in central Arkansas with potential for damaging earthquake. *Seismological Research Letters* 83(2):250-260.

Howe, A.M., C.T. Hayward, B.W. Stump, and C. Frohlich. 2010. Analysis of recent earthquakes in Cleburne, Texas (Abstract). *Seismological Research Letters* 81:379.

Julian, B.R., G.R. Fouilger, and F. Monastero. 2007. Microearthquake moment tensors from the Coso geothermal field. Proceedings, 32nd Workshop on Geothermal Reservoir Engineering, Stanford University, CA, January 22-24.

Kaieda, H., S. Shunji Sasaki, and D. Wyborn. 2010. Comparison of characteristics of micro-earthquakes observed during hydraulic stimulation operations in Ogachi, Hijiori and Cooper Basin HDR projects. Proceedings, World Geothermal Congress, Bali, Indonesia, April 25-29.

Keller, G.R., A.M. Robers, and C.D. Orr. 1987. Seismic activity in the Permian Basin area of west Texas and southeastern New Mexico, 1975-1979. *Seismological Research Letters* 58(2):63-70.

Kouznetsov, O., V. Sidorov, S. Katz, and G. Chilingarian. 1994. Interrelationships among seismic and short-term tectonic activity, oil and gas production, and gas migration to the surface. *Journal of Petroleum Science and Engineering* 13:57-63.

Kovach, R.L. 1974. Source mechanisms for Wilmington oil field, California, subsidence earthquakes. *Bulletin of the Seismological Society of America* 64(3):699-711.

Kugaenko, Y., V. Saltykov, and V. Chebrov. 2005. Seismic situation and necessity of local seismic monitoring in exploited Mutnovsky steam-hydrothermal field (southern Kamchatka, Russia). Proceedings, World Geothermal Congress, Antalya, Turkey, April 24-29.

Lei, Z., G. Yu, S. Ma, X. Wen, and Q. Wang. 2008. Earthquakes induced by water injection at ~3 km depth within the Rongchang gas field, Chongqing, China. *Journal of Geophysical Research* 113:B10310.

Majer, E.L., R. Baria, M. Stark, S. Oates, J. Bommer, B. Smith, and H. Asanuma. 2007. Induced seismicity associated with enhanced geothermal systems. *Geothermics* 36(3):185-222.

McGarr, A. 1991. On a possible connection between three major earthquakes in California and oil production. *Bulletin of the Seismological Society of America* 81(3):948-970.

Nicholson, C., and R.L. Wesson. 1990. Earthquake Hazard Associated with Deep Well Injection—A Report to the US Environmental Protection Agency. U.S. Geological Survey Bulletin 1951, 74 pp.

Nicholson, C., and R.L. Wesson. 1992. Triggered earthquakes and deep well activities. *Pure and Applied Geophysics* 139(3/4):562-578.

Ohtake, M. 1974. Seismic activity induced by water injection at Matsushiro, Japan. *Journal of Physics of the Earth* 22(1):163-176.

Pennington, W.D., S.D. Davis, S.M. Carlson, J. DuPree, and T.E. Ewing. 1986. The evolution of seismic barriers and asperities caused by the depressuring of fault planes in oil and gas fields of south Texas. *Bulletin of the Seismological Society of America* 76(4):939-948.

Rothe, G.H., and C.-Y. Lui. 1983. Possibility of induced seismicity in the vicinity of the Sleepy Hollow oil field, southwestern Nebraska. *Bulletin of the Seismological Society of America* 73(5):1357-1367.

Shouzhong, D., Z. Huanpeng, and G. Aixiang. 1987. Rare seismic clusters induced by water injection in the Jiao well 07 in Shengli oil field. *Earthquake Research in China* 1:313.

Smirnova, M.N. 1968. Effect of earthquakes on the oil yield of the Gudermes field (northeastern Caucasus). *Izvestiya, Earth Physics* 12:760-763.

Sze, E.K.-M. 2005. Induced seismicity analysis for reservoir characterization at a petroleum field in Oman. Ph.D. thesis, Massachusetts Institute of Technology, Cambridge.

van Eck, T., F. Goutbeek, H. Haak, and B. Dost. 2006. Seismic hazard due to small-magnitude, shallow-source, induce earthquakes in the Netherlands. *Engineering Geology* 87(1-2):105-121.

Wetmiller, R.J. 1986. Earthquakes near Rocky Mountain House, Alberta and their relationship to gas production facilities. *Canadian Journal of Earth Sciences* 23. 172 pp.

Zoback, M.D., and J.C. Zinke. 2002. Production-induced normal faulting in the Valhall and Ekofisk oil fields. *Pure and Applied Geophysics* 159:403-420.

Letters between Senator Bingaman and Secretary Chu

Syec-2010-010389

United States Senate

COMMITTEE ON
ENERGY AND NATURAL RESOURCES
WASHINGTON, DC 20510-6150

ENERGY.SENATE.GOV

June 17, 2010

The Honorable Steven Chu
Secretary of Energy
U.S. Department of Energy
1000 Independence Avenue, SW
Washington, DC 20585

Dear Mr. Secretary:

Many of the next generation energy technologies vital for our country's future require the injecting of fluids – be they water, carbon dioxide, or other mixes – deep into the earth's subsurface. Geothermal energy extraction, geologic carbon sequestration, hydraulic fracturing to extract natural gas from shales, and enhanced oil recovery all require the injection and movement of fluids deep underground, a process that by its very nature may induce seismic activity. I understand that the Department of Energy has recently initiated studies in several of its offices and programs to address the issue of induced seismicity, and I commend those efforts.

I am writing to ask that the Department of Energy, in cooperation with the Department of the Interior and all other relevant agencies, initiate a comprehensive and independent National Academy of Sciences and the National Academy of Engineering study to examine the possible scale, scope, and consequences of seismicity induced by energy technologies. Though oil and natural gas extraction processes have moved fluids through the ground for many decades without significant seismic consequences, the prospect of greatly increased deployment of these new energy technologies in the coming years, coupled with a commensurate rising public concern about their safety, makes it necessary to now better understand the nature and scale of seismicity that may be induced by all subsurface energy activity.

Recent studies such as the 2010 joint University of Texas – Southern Methodist University article by Cliff Frohlich et al. regarding the correlations of seismic activity with natural gas extraction activities in Texas, and the 2007 study led by Ernest L. Majer of the Lawrence Berkeley National Laboratory entitled, "Induced Seismicity Associated with Enhanced Geothermal Systems," indicate a possible link between energy-related subsurface fluid movement and increased seismic activity. Importantly, both studies found that all recorded earthquakes that may have been induced by energy projects were small (less than 4.6 on the Richter scale) and had few or no significant impacts on human health or property. However, both studies emphasize that a more extensive, thorough, and definitive study is necessary to fill gaps in existing knowledge, such as how subsurface energy activities interact with existing geologic stresses to increase or decrease the risk of induced seismic events.

Such a comprehensive study – conducted by the scientifically trusted, nationally recognized, and independent National Academies – will give policymakers the information they need to develop better safety guidelines and regulations for these important energy technologies. It will also provide energy developers with tools to implement appropriate risk mitigation efforts and to choose safe sites for new projects, and arm the public with the information they need to be confident in the safety of their homes and families.

Much of public opposition to the deployment of advanced energy technologies in the United States stems from a lack of clear, trusted information regarding the safety of those new energy facilities for the local communities that are their neighbors. A National Academies study can provide information to these concerned communities – whether near a new geothermal facility tapping heat trapped deep in the earth, a carbon sequestration site storing carbon dioxide underground to facilitate a new clean coal future, a drill rig extracting the newfound riches of America's shale gas, or an aging domestic oil well rejuvenated by enhanced recovery techniques that replaces foreign oil with domestic production – and allow America to proceed safely and with confidence to a cleaner and more secure energy future.

I appreciate your consideration of this request, and look forward to working with you on this.

Sincerely,

Jeff Bingaman
Chairman

EXEC-2010-010389

Department of Energy
Washington, DC 20585

June 24, 2010

The Honorable Jeff Bingaman
United States Senate
Washington, DC 20510

Dear Senator Bingaman:

Thank you for your June 17, 2010, letter to Secretary Chu asking that the Department of Energy (DOE), in cooperation with the Department of the Interior and all other relevant agencies, initiate a comprehensive and independent National Academy of Sciences and National Academy of Engineering study to examine the possible scale, scope, and consequences of seismicity induced by energy technologies. As you noted, geothermal energy extraction, geologic carbon sequestration, hydraulic fracturing to extract natural gas from shales, and enhanced oil recovery all require the injection and movement of fluids deep underground that may induce seismic activity. DOE has initiated studies in several program areas to address the issue of induced seismicity, but more extensive, thorough, and definitive study is warranted to fill gaps in existing knowledge.

A comprehensive and independent National Academy of Sciences and National Academy of Engineering study of this subject should be undertaken. Enhanced oil recovery and other processes have required movement of fluids deep underground for many years without significant seismic consequences. However, future increased deployment of new energy technologies, combined with possible increased public concern about safety issues, make an independent study appropriate to provide a better understanding of the nature and scale of seismicity that may be induced by all subsurface energy activity.

DOE, in cooperation with the Department of the Interior and other appropriate agencies, will initiate activities to proceed with a comprehensive study, to be conducted by the National Academies, that will give policymakers the information necessary to develop safety guidelines and regulations for these important energy technologies. The resulting study will also provide energy developers with information needed for risk mitigation efforts and to choose safe sites for new projects.

It is understood that public opposition to the deployment of advanced energy technologies in the United States derives, in part, from a perceived lack of trusted information regarding the safety of those new energy facilities by their local neighborhood communities. You can rest assured that the Department will continue its efforts to provide the most accurate, trusted information possible regarding the safety of those new energy facilities to the public. The proposed

National Academies study concerning possible seismicity induced by energy technologies can provide reliable information to these concerned communities and allow America to proceed safely to a cleaner and more secure energy future.

I appreciate your interest and look forward to working with you on this issue. Please do not hesitate to contact me or Elizabeth Nolan, Office of Congressional and Intergovernmental Affairs, 202-586-5450, if you have any further questions.

Sincerely,

James J. Markowsky
Assistant Secretary
Office of Fossil Energy

Earthquake Size Estimates and Negative Earthquake Magnitudes

The original and arguably the best-known magnitude scale for measuring the size of an earthquake is the Richter scale, derived by Charles Richter in 1935 at the California Institute of Technology to measure earthquake size in Southern California. Using an early seismograph he defined local magnitude M_L to be

$$M_L = \mathrm{Log}A - \mathrm{Log}A_o$$

where A is the maximum amplitude of deflection of a needle on a chart, in millimeters, measured on the seismograph. A_o is an empirical distance correction appropriate for the region (Richter, 1936). Richter assigned a magnitude 3 to an event with amplitude of 1 mm recorded on a Wood Anderson seismograph at 100 km distance from the source, and a magnitude 0 with amplitude 0.001 mm at 100 km, thought to be the smallest possible instrumentally recorded earthquake (Shemeta, 2010).

Since the 1930s advancements in equipment design such as more sensitive geophones and digital recording equipment and closer proximity to earthquake sources dramatically advanced the ability to record and analyze data from small earthquakes. Using borehole seismic arrays located within a few hundred meters of an earthquake source, very small earthquakes can be recorded. These events are smaller than the baseline magnitude of "0" originally designed by Richter, therefore the range of event sizes continues into the negative magnitude range (Figure E.1).

Because the Richter scale was designed for the Wood Anderson seismograph measurements, its routine use in modern seismology is now quite limited; however, most modern earthquake magnitudes are based on scales that relate back to the Richter scale.

OTHER SIZE ESTIMATES FOR EARTHQUAKES

In practice Richter's method for estimating earthquake magnitude has been largely supplanted by other more flexible and robust measures of magnitude. The moment magnitude, which is scaled to agree with the Richter magnitude, is in wide use because it can be

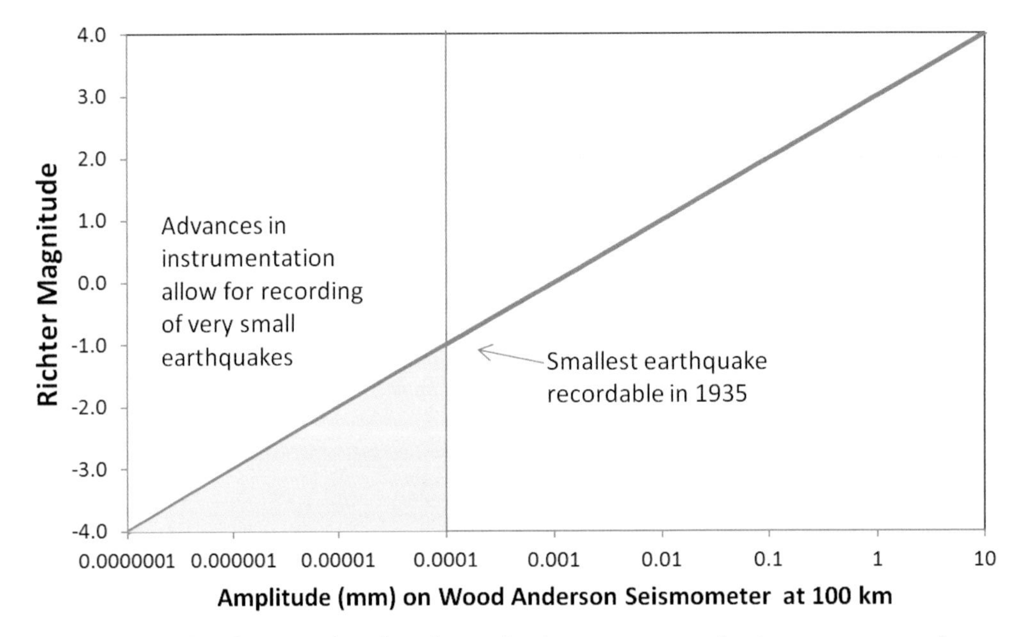

FIGURE E.1 A plot of measured earthquake amplitude versus magnitude. The more sensitive the seismic instruments, the smaller the measureable magnitude, reaching into the negative magnitude range.

tied to other direct measures of the size of an earthquake. The seismic moment is a routine measurement describing the strength of an earthquake and is defined as

$$M_o = \mu S d$$

where **μ** is the shear modulus, **S** is the surface area of the fault, and **d** is the average displacement along the fault. The moment magnitude, M_w, is related to seismic moment by the Hanks and Kanamori (1979) equation

$$M_w = {}^{2}/_{3} \, \mathrm{Log} M_o - 6$$

where M_o is in Newton meters, valid for earthquakes ranging from magnitude 3 to 7 (Shemeta, 2010). There are a variety of methods used to calculate a seismic moment from microseismic waveforms.

EARTHQUAKE "B VALUES"

Small earthquakes occur much more often than large earthquakes. The number of earthquakes with respect to magnitude follows a power law distribution and is described by

$$\text{Log}_{10}\mathbf{N} = \mathbf{a} - \mathbf{bM}$$

where **N** is the cumulative number of earthquakes with magnitudes equal to or larger than **M**, and **a** is the number of events of **M** = 0. The variable **b** describes the relationship between the number of large and small events and is the slope of the best-fit line between the number of earthquakes at a given magnitude and the magnitude (Gutenberg and Richter, 1944; Ishimoto and Iida, 1939). A **b** value close to 1.0 is commonly observed in many parts of the world for tectonic earthquakes. This relationship is often referred to as the Gutenberg-Richter magnitude frequency relationship.

Differences in the slope **b** reveal information about the potential size and expected number of the events in a population of earthquakes. Analysis of **b** values around the world has shown that in fluid injection scenarios the **b** value is often in the range of 2, which reflects a larger number of small events (swarm earthquakes), compared to tectonic earthquakes. In hydraulic fracturing microseismicity, **b** values in the range of 2 are commonly observed (Maxwell et al., 2008; Urbancic et al., 2010; Wessels et al., 2011). The high **b** values observed in hydraulic fracturing are thought to represent the opening of numerous small natural fractures during the high-pressure injection (Figure E.2). It is possible for a hydraulic fracture to grow into a nearby fault and reactivate it, if the orientation of the fault is favorable for slip under the current stress conditions in the reservoir. Figure E.3 is an example of a hydraulic fracture reactivating a small fault during injection.

REFERENCES

Gutenberg, B., and C.F. Richter. 1944. Frequency of earthquakes in California. *Bulletin of the Seismological Society of America* 34:185-188.

Hanks, T.C., and H. Kanamori. 1979. A moment magnitude scale. *Journal of Geophysical Research* 84(B5):2348-2350.

Ishimoto, M., and K. Iida. 1939. Observations of earthquakes registered with the microseismograph constructed recently. *Bulletin of the Earthquake Research Institute* 17:443-478.

Maxwell, S.C., J. Shemeta, E. Campbell, and D. Quirk. 2008. Microseismic deformation rate monitoring. Society of Petroleum Engineers (SPE) 116596-MS. SPE Annual Technical Conference and Exhibition, Denver, Colorado, September 21-24.

Richter, C.F. 1936. An instrumental earthquake magnitude scale. *Bulletin of the Seismological Society of America* 25:1-32.

Shemeta, J. 2010. It's a matter of size: Magnitude and moment estimates for microseismic data. *The Leading Edge* 29(3):296.

Urbancic, T., A. Baig, and S. Bowman. 2010. Utilizing b-values and Fractal Dimension for Characterizing Hydraulic Fracture Complexity. GeoCanada—Working with the Earth. ESG Solutions. Available at www.geocanada2010.ca/uploads/abstracts_new/view.php?item_id=976 (accessed April 2012).

Wessels, S.A., A. De La Pena, M. Kratz, S. Williams-Stroud, and T. Jbeili. 2011. Identifying faults and fractures in unconventional reservoirs through microseismic monitoring. *First Break* 29(7):99-104.

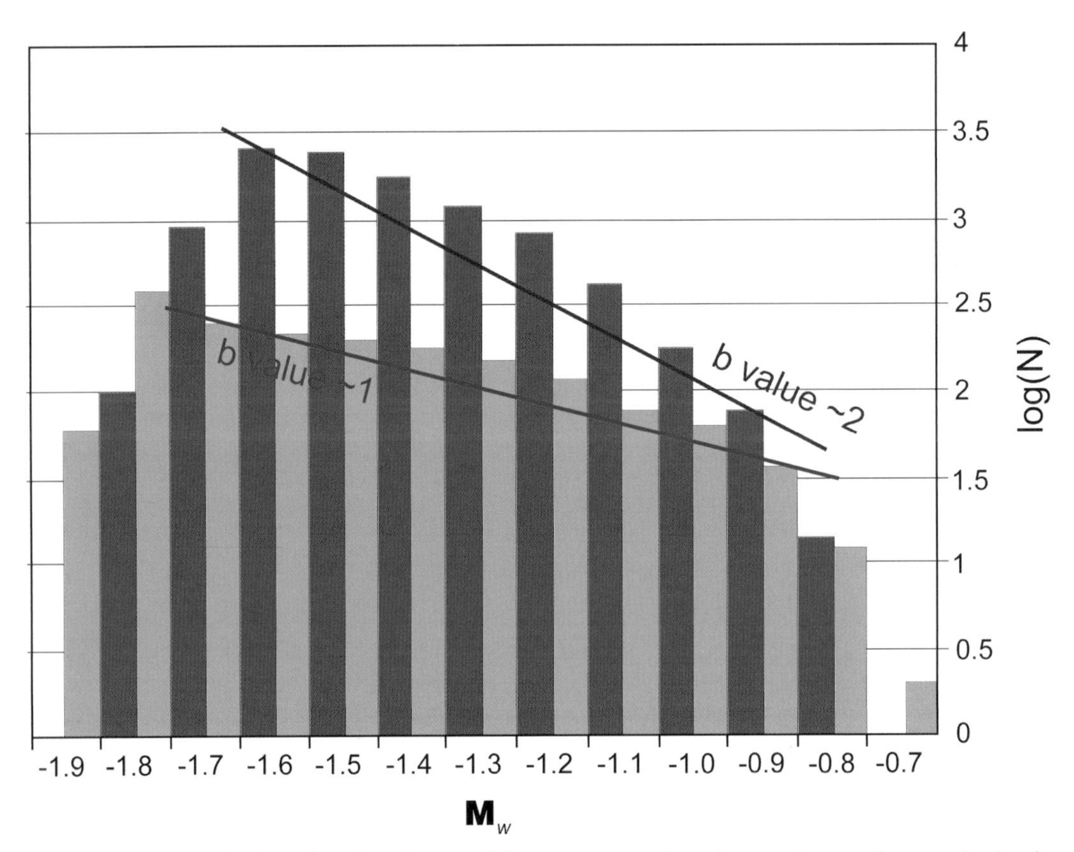

FIGURE E.2 Graph shows b values for two different microearthquake populations during a hydraulic fracture treatment. The b values vary from about 1 for reactivated tectonic microseismic events and 2 for microseismicity associated with the hydraulic fracture injection. The hydraulic fracture microseismic magnitudes are typically very small (less than **M** 0), hence the lack of larger microseismic events on this b value example. SOURCE: From Wessels et al. (2011).

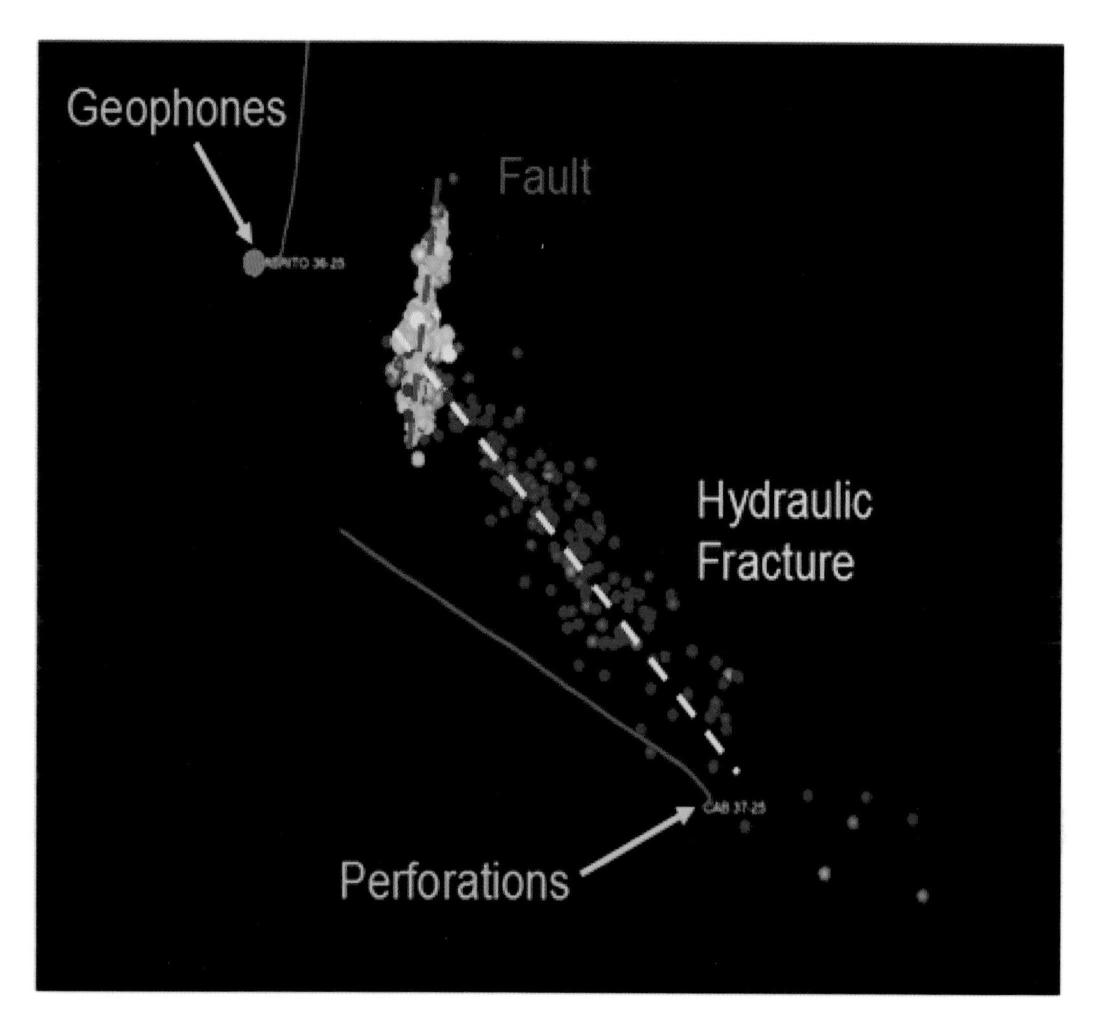

FIGURE E.3 Example of a reactivated fault during hydraulic fracturing. The figure is a map view of a microseismicity (colored spheres which are colored by magnitude; cool colors are small events) during a hydraulic fracture treatment. The fracturing well is shown by the pink line and is deviated away from a central wellhead location and extends vertically through the reservoir section; the injection location is labeled "Perforations." The data were recorded and analyzed using borehole receivers (marked Geophones). The blue dots show the growth of the hydraulic fracture to the northwest, then intersecting and reactivating a small fault in the reservoir, shown by change in fracture orientation and larger magnitude events (yellow dots). SOURCE: From Maxwell et al. (2008).

The Failure of the Baldwin Hills Reservoir Dam

On December 14, 1963, the dam built to contain the Baldwin Hill Reservoir located in southwest Los Angeles failed, releasing 250 million gallons of water into the housing subdivisions below the dam. Approximately 277 homes were damaged or destroyed and five people were killed by the disaster (Hamilton and Meehan, 1971). Although there is speculation that waterflooding operations in the Inglewood Oil Field (located to the west and south of the reservoir) were partially to blame for the failure of the reservoir dam, the dam itself did not fail due to an induced earthquake. Records from the Seismographic Laboratory of the California Institute of Technology located 15 miles northeast of the reservoir showed no earthquakes large enough to cause internal damage to the reservoir during the period 1950-1963 (Jansen, 1988). Instead, the sealing layers in the floor of the reservoir failed due to the "creep" of several geologic fractures below the reservoir, which caused the release of water through the floor of the reservoir that resulted in the structural failure of the dam itself.

The Baldwin Hills Reservoir was constructed between 1947 and 1951 by the Los Angeles Department of Water and Power. The reservoir was constructed on a hilltop and was formed by a dam on the north side and earthen dikes on the other three sides, which were constructed of materials excavated from the reservoir bowl. The soil under the reservoir was composed of porous material and was bisected by three known geologic faults (Jansen, 1988). The floor of the reservoir was made watertight by the use of two layers of asphalt with compacted earth between them. Below the upper layer of asphalt and earth, a level of pea gravel with tile drains was installed to allow the monitoring of leakage from the bottom of the reservoir. Extensive discharge from the drainage system was recorded during the initial filling of the reservoir, and filling was discontinued until repairs to the reservoir could be made (Jansen, 1988). Cracking in concrete portions of the reservoir was noted as early as 1951.

The Inglewood Oil Field was discovered in 1924 and covered approximately 1,200 acres when fully developed. At the time of the failure of Baldwin Hills Dam in 1963, the field had more than 600 producing wells, and the closest wells were located within 700 feet of the reservoir structure. The oil reservoir is divided into multiple compartments due to a series of geologic faults. Several of these faults not only divide the Inglewood Oil Field but also continue to the surface and are present on the site of the Baldwin Hills Reservoir. The depth of the wells in the Inglewood Field is between 2,000 and 4,000 feet. Due to subsurface fluid withdrawal, the ground level above the field exhibited a surface subsidence of approximately

10 feet by 1964. In order to increase production, waterflooding operations were commenced in 1954 and expanded in 1955 and 1961. These injection operations increased pore pressure in portions of the oil field from 50 psi to over 850 psi by 1963 (Hamilton and Meehan, 1971). Injection depths were as shallow as 1,200 feet.

The dam structure failed due to subsurface leakage of reservoir water beneath the floor of the impoundment and under the foundation of the dam itself. The subsurface leakage was caused by a cracked seal extending across the floor of the reservoir in line with the breach in the dam (Jansen, 1988). Movement of the geologic faults crossing the floor of the reservoir with downward displacement of 2 to 7 inches on the western side of several faults caused cracking in the asphalt membrane seal and allowed water to enter the porous soil beneath the dam. Later excavations of the bottom of the reservoir indicated that leakage had occurred for an appreciable amount of time before the dam failure. The slow movement of the faults beneath the reservoir has been attributed to (1) natural causes inherent in the geologic setting, (2) subsidence of the ground surface caused by oil and gas operations or by the filling of the reservoir with water, or (3) pressure injection of water in the Inglewood Field at shallow depths for oil and gas operations and in the presence of a fault system.

REFERENCE

Hamilton, D.H., and R.L. Meehan. 1971. Ground rupture in the Baldwin Hills. *Science* 172(3981):326-406.
Jansen, R.B. 1988. *Advanced Dam Engineering for Design, Construction, and Rehabilitation*. New York: Springer.

Seismic Event Due to Fluid Injection or Withdrawal

To initiate a seismic event by activation of an existing fault, a critical condition involving the in situ state of stress and the pore pressure needs to be met. As discussed below, this condition stems, at least for the simplest case of slip initiation along a preexisting fault, from a combination of two fundamental concepts: (1) slip is initiated when the shear stress acting on the fault overcomes the frictional resistance and (2) the frictional resistance is given by the product of the friction coefficient times the normal effective stress, defined as the normal stress across the fault reduced by the fluid pressure. This condition of slip initiation, referred to as the Coulomb criterion, can then be translated as a limit condition on the magnitude of the vertical and horizontal stress and of the pore pressure, which depends on the inclination of the fault. The formation of a fault follows similar concepts but accounts for an additional shear resistance due to cohesion; also the actual orientation of the created fault corresponds to the inclination for which the condition of slip is first met.

Although the initial in situ stress state and pore pressure are often close to the limit condition required to cause slip on an existing fault, not all perturbations in the stress and pore pressure associated with fluid injection or extraction eventually trigger a seismic event. First, the perturbation must be destabilizing in its nature; that is, it must bring the system closer to critical conditions, irrespective of the magnitude of the perturbation. Indeed some perturbations are stabilizing, meaning that they move the system farther away from critical conditions. The degree of destabilization can be assessed by a certain parameter m that characterizes the nature of the stress and pore pressure perturbation (Figure G.1). Second, if the perturbation is indeed destabilizing, the magnitude of the perturbation has to be large enough to reach critical conditions. Finally, not all slip events are seismic, although most are, as gouge-filled faults could respond in a ductile stable manner.

It is useful to contrast the case of fluid injection in reservoir rocks, where the fluid flows and is stored in the pore network of the rock, from that in crystalline impermeable rocks, where the injected fluid is essentially transmitted and stored in the fracture network. In the permeable case, the pore pressure increases in the rock induce stress variation in the reservoir and in the surrounding rock. In the impermeable case, the stress induced by injection is negligible (except in situations where the fracture network is very dense), but fluid pressure change can be transmitted over a large distance by fractures that offer little resistance to flow. Although our analysis in this appendix refers to a finite-extent reservoir, solution of the infinite case lies within the finite solution. For the purposes of understanding pore

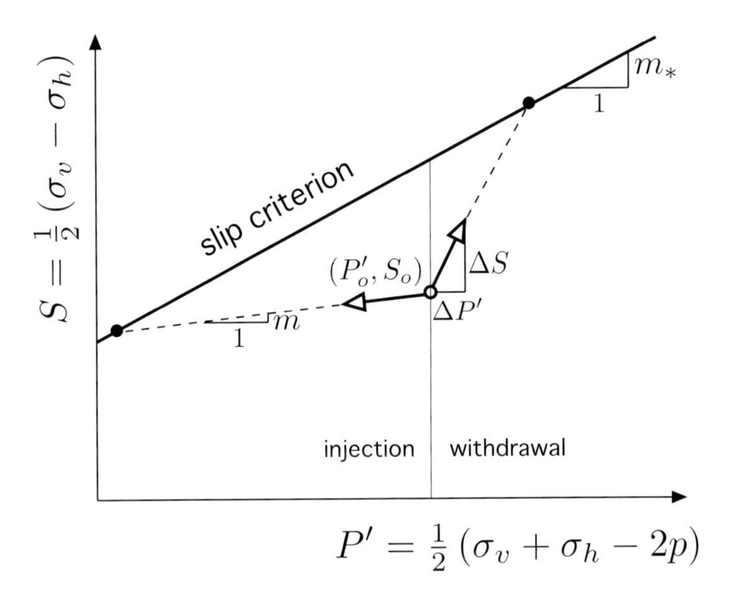

FIGURE G.1 Effective stress change in a reservoir induced by injection or withdrawal of fluid.

pressure perturbation in an infinite reservoir, one simply takes the length of the reservoir to infinity, which causes the reference time scale to go to infinity.

FLUID INJECTION AND EXTRACTION IN A (PERMEABLE) RESERVOIR ROCK

An increase of pore pressure in a permeable rock that is free to deform induces an increase of volume. This physical phenomenon is akin to thermal expansion (i.e., the volume increase experienced by an unconstrained material when subjected to a temperature increase). However, because the deformation of the rock is inhibited by the surrounding material, an increase of pore pressure induces a volume change that is smaller than the unconstrained volume change that would have been for the same pore pressure increase. In addition the compressive stresses in the rock are increased by an amount proportional to the pore pressure increase (see Box 2.3). But for very specific situations, the compressive stress increases in the vertical and in the horizontal directions are unequal, the stress ratio being a function of the shape of the reservoir and the contrast in elastic properties between the reservoir and the surrounding rocks (Rudnicki, 1999, 2002). In particular, the ratio of the induced vertical stress to the induced horizontal stress decreases with the aspect ratio of the reservoir (i.e., the ratio of the reservoir thickness to the lateral extent). For a "thin" reservoir, characterized by a small aspect ratio, the vertical stress change is negligible, and all the stress increase takes place in the horizontal direction, with increases that range between 40 and 80 percent of the pore pressure increase.

The expansion of the reservoir as a whole also alters the stress state in the surrounding rock, in particular inducing a decrease of the horizontal stress above and below a thin reservoir. These stress variations could in principle also trigger normal faulting in these regions; however, the combination of stress and pore pressure change caused by fluid injection is more likely to trigger seismicity in the reservoir rather than outside. The reverse is true for fluid extraction.

FLUID INJECTION IN A FRACTURED IMPERMEABLE ROCK

Unlike fluid injection in permeable rocks, the injection of fluid in fractured impermeable rock is essentially inducing an increase of fluid pressure in the fractures, with negligible concomitant changes in the stress. It is therefore a worst case compared to the permeable rock case, where the increase of pore pressure is in part offset by an increase of the compressive stress, which is a stabilizing factor. (In other words, factor m introduced in Figure G.1 is about equal to zero.) Because fractures can be very conductive and offer less storage compared to a permeable rock, the pore pressure perturbations can travel on the order of kilometers from the point of injection.

Coulomb Criterion and Effective Stress

For slip to take place on a fault, a critical condition involving the normal stress σ (the force per unit area normal to the fault), the shear stress τ (the force per unit area parallel to the fault), and the pressure ρ of the fluid on the fault plane, must be met (see Figure G.2 for a representation of σ and τ). This condition is embodied in the Coulomb criterion, $|\tau| = \mu(\sigma - \rho) + c$, which depends on two parameters: the coefficient of friction μ, with values typically in the narrow range from 0.6 to 0.8, and the cohesion c, equal to zero, however, for a frictional fault.

The Coulomb criterion simply expresses that the condition for slip on the fault is met when the magnitude of the "driving" shear stress, $|\tau|$, is equal to the shear resistance $\mu(\sigma - \rho) + c$. The quantity $(\sigma - \rho)$ is known as the effective stress, a concept initially introduced by Terzaghi (1940) in the context of soil failure. It captures the counteracting influence of the fluid pressure ρ on the fault to the stabilizing effect of the compressive stress σ acting across the fault.

As long as the shear resistance is larger than the shear stress magnitude, the fault is stable. However, an increase of the shear stress magnitude or a decrease of the shear strength would cause the fault to slip if the two quantities become equal. For example, an increase of the fluid pressure induced by injection could be responsible for a drop of shear strength large enough to reach the critical conditions.

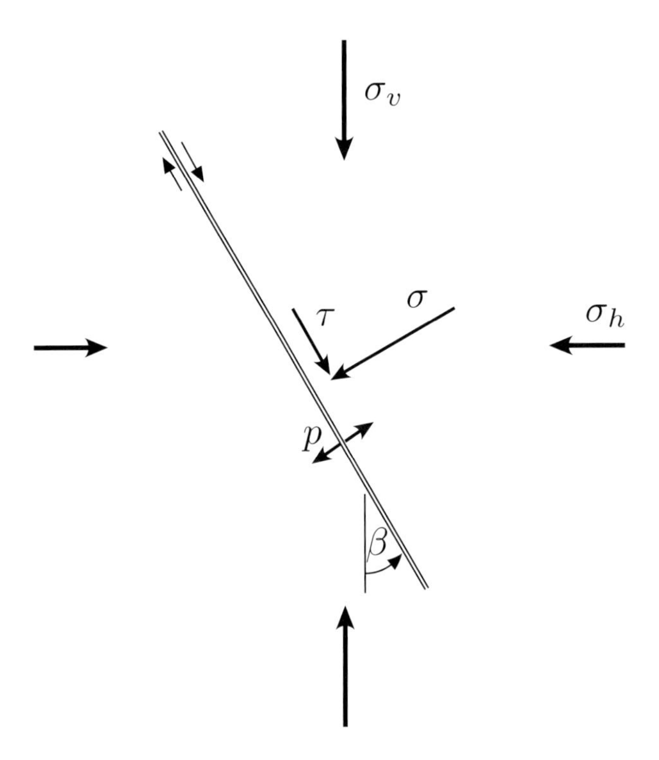

FIGURE G.2 The normal and shear stress, σ and τ, acting across the fault depends on the vertical and horizontal stresses, σ_v and σ_h, and the fault inclination β. The fault is infiltrated by fluid at pressure ρ.

The normal and shear stress on the fault can actually be expressed in terms of the in situ vertical and horizontal stresses, σ_v and σ_h, through a relation that depends on the fault inclination β (Figure G.2). The above Coulomb criterion can then be expressed as a limiting condition in terms of the effective vertical and horizontal stresses $\sigma'_v = \sigma_v - \rho$ and $\sigma'_h = \sigma_h - \rho$ or equivalently in terms of their half-sum and half-difference, P' and S. Figure G.3 provides a graphical representation of the Coulomb criterion in terms of these two quantities.

The fault is stable if the point representative of the (effective) in situ stress state is below the slip criterion. A perturbation ($\Delta P'$, ΔS), induced by fluid injection or withdrawal, to an existing state (P'_o, S_o) that moves the point ($P'_o + \Delta P'$, $S_o + \Delta S$) to be on the Coulomb line will cause slip and trigger a seismic event. However, only some perturbations are destabilizing in nature (i.e., they move the representative stress point [P', S] closer to the critical conditions). For example, the destabilizing perturbation shown in Figure G.3 is characterized by a slope $m = \Delta S/\Delta P'$ smaller than m_o and a "direction" corresponding to both $\Delta P'$ and ΔS being negative. A perturbation characterized by the same slope m, but positive variations $\Delta P'$ and ΔS, would be stabilizing.

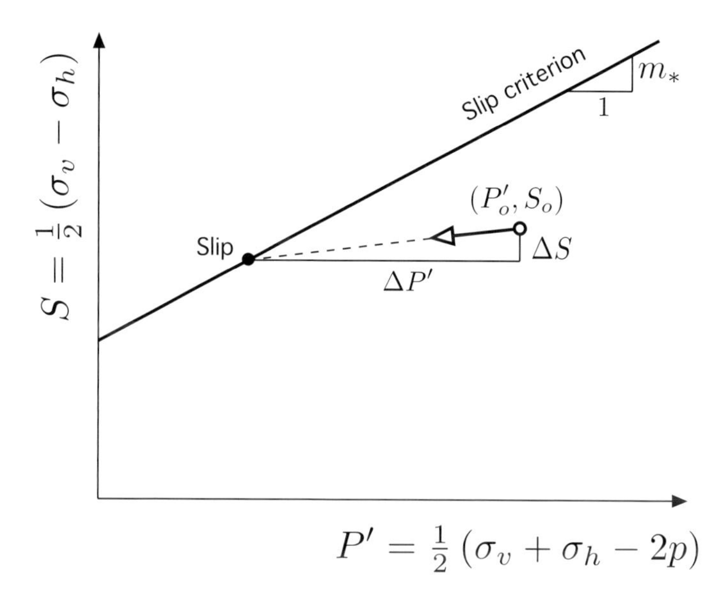

FIGURE G.3 Stress and pore pressure perturbations from an initial stable state leading to critical conditions. The vertical intercept represents the rock cohesive strength and is zero for a preexisting frictional fault. The slope m_o of the slip criterion depends on the friction coefficient μ and on the fault inclination β. The sketch corresponds to the normal conditions when $\sigma'_v > \sigma'_h$.

The existence of a perturbation ΔS reflects the fact that injection or extraction of fluid in deep layers has consequences beyond simply increasing or decreasing the pore fluid pressure. As explained in Chapter 2, the propensity of permeable rocks to expand (contract) as a response to increase (decrease) of pore pressure induces stress change not only in the reservoir but also in the surrounding rocks. Only in the particular case of impermeable rocks, where flow of fluids only takes place in a fracture network, are the perturbations essentially only of a hydraulic nature. For example, injection of fluid in fractured impermeable rock causes mainly an increase of pore pressure Δρ leading to $\Delta P' < 0$ and $\Delta S = 0$, which would cause the stress point in Figure G.3 to move horizontally ($m = 0$) to the left.

So far the discussion has been focused on slip on a preexisting fault of known inclination β. The formation of a fault associated with the large-scale shear failure of the rock can be treated within the same framework, with the critical difference that the inclination of the created fault depends only on the friction coefficient μ. It also follows that in the representation of Figure G.3, the slope m_o of the slip criterion (now usually referred to as the Mohr-Coulomb criterion) is exclusively a function of μ. The vertical intercept of the Mohr-Coulomb criterion with the *S* axis then embodies the cohesive shear strength of the rock.

REFERENCES

Rudnicki, J.W. 1999. Alteration of regional stress by reservoirs and other inhomogeneities: Stabilizing or destabilizing? Pp. 1629-1637 in *Proceedings of the Ninth International Congress on Rock Mechanics*, edited by G. Vouille and P. Berest. London: Taylor & Francis.

Rudnicki, J.W. 2002. Eshelby transformations, pore pressure and fluid mass changes, and subsidence. Pp. 307-312 in *Poromechanics II, Proceedings of the Second Biot Conference on Poromechanics*, edited by J.-L. Auriault et al. Leiden: A.A. Balkema.

Terzaghi, K. 1940. *Theoretical Soil Mechanics*. New York: Wiley.

Pore Pressure Induced by Fluid Injection

The dependence of the induced pore pressure on the operation parameters (injection rate, volume of fluid injected), on position and time, and on the hydraulic properties of the reservoir is illustrated in this appendix by considering the simple example of fluid injection in a disk-shaped reservoir. The analysis shows that different parameters control the pore pressure at the beginning of the injection operation and once enough fluid has been injected in the reservoir (see also Nicholson and Wesson, 1990).

The pore pressure induced by injection of fluid, $\Delta\rho$, is to a good approximation governed by the diffusion equation

$$c\nabla^2\Delta\rho = \partial\Delta\rho/\partial t + \text{source}$$

where c denotes the hydraulic diffusivity equal to $c = k/\mu S$. In the above, k is the intrinsic permeability of the rock (generally expressed in Darcy), μ is the fluid viscosity, and S is the storage coefficient, a function of the compressibility of both the fluid and the porous rock. The diffusion equation imposes a certain structure on the link between the magnitude of the induced pore pressure $\Delta\rho$, the injected fluid volume V, and the rate of injection Q_o.

As an example, we consider the injection of fluid at a constant volumetric rate Q_o, at the center of a disk-shaped reservoir of thickness H and radius R. It is assumed that the reservoir is thin (i.e., $H/R \ll 1$), and also that the pore pressure is uniform over the thickness of the layer, which implies, depending on the manner the fluid is injected, that some time has elapsed since the beginning of the operation.

At early time (to be defined more precisely later), the pore pressure perturbation induced by injection of fluid has not reached the boundary of the reservoir. The induced pore pressure field is then given by the source solution for an infinite domain, a solution of the form (Wang, 2000)

$$\Delta\rho(r,t) = \rho_* F(r/\sqrt{[ct]}) \qquad (1)$$

where r is the radial distance from the injection well, t is time, and F is a known function. The quantity where ρ_* is a characteristic pressure (i.e., a yardstick for measuring the induced pressure) given by

$$\rho_* = \mu Q_o/kH$$

Once the time elapsed since injection started becomes larger than a fraction, say 0.1, of the characteristic time $t_* = R^2/c$, then the evolution of the induced pore pressure becomes influenced by the finiteness of the reservoir. Formally, the pore pressure solution can then be expressed as

$$\Delta p(r,t) = p_* P(r/R, t/t_*) \quad (2)$$

The function $P(\rho,t)$ can be determined semianalytically. If the elapsed time t is expressed as the ratio of the injected volume V to the rate of injection Q_o (i.e., $t = V/Q_o$), then solution (2) can be written as

$$\Delta p(r,V) = p_* P(r/R, V/V_*) \quad (3)$$

where $V_* = (Q_o R^2)/c$ is a characteristic fluid volume. The above expression suggests that the relationship between the induced pore pressure Δp, the injected volume V, and the injection rate Q_o is not straightforward. However, Equation (3) shows important trends; for example, a decrease of the permeability causes an increase of the characteristic pressure, or an increase of the storage coefficient causes a decrease of the pore pressure, all other parameters kept constant.

At small time $t \ll t_*$, the dimensionless pressure $P = \Delta p/p_*$ reduces to the unbounded domain solution F, while at large time $t \gg t_*$, the pressure tends to become uniform and the pore pressure is simply given by

$$p \cong V/(\pi R^2 HS) \quad (4)$$

as the function $P(\rho,t)$ behaves for large t as $P \cong t/\pi$. Thus, at large time, the pore pressure is simply proportional to the volume of injected fluid (Figure H.1). Equation (4) actually indicates that the large-time pore pressure is simply the ratio of the injected volume over the reservoir volume, divided by the storage coefficient.

The previous material provides some information about the link between pore pressure, injected volume, and injected rate for the particular case of an injector centered in a disk-shaped reservoir. These ideas can be generalized to more realistic cases. For example, for an arbitrarily shaped reservoir with n wells, each injecting at a rate Q_o, the general expression for the induced pore pressure can be written as

$$\Delta p(x,t) = p_* \varsigma\{x/L, t/t_*; n, (x_i, i=1, n), \textit{reservoir shape}\}$$

where the characteristic pressure and time are given by

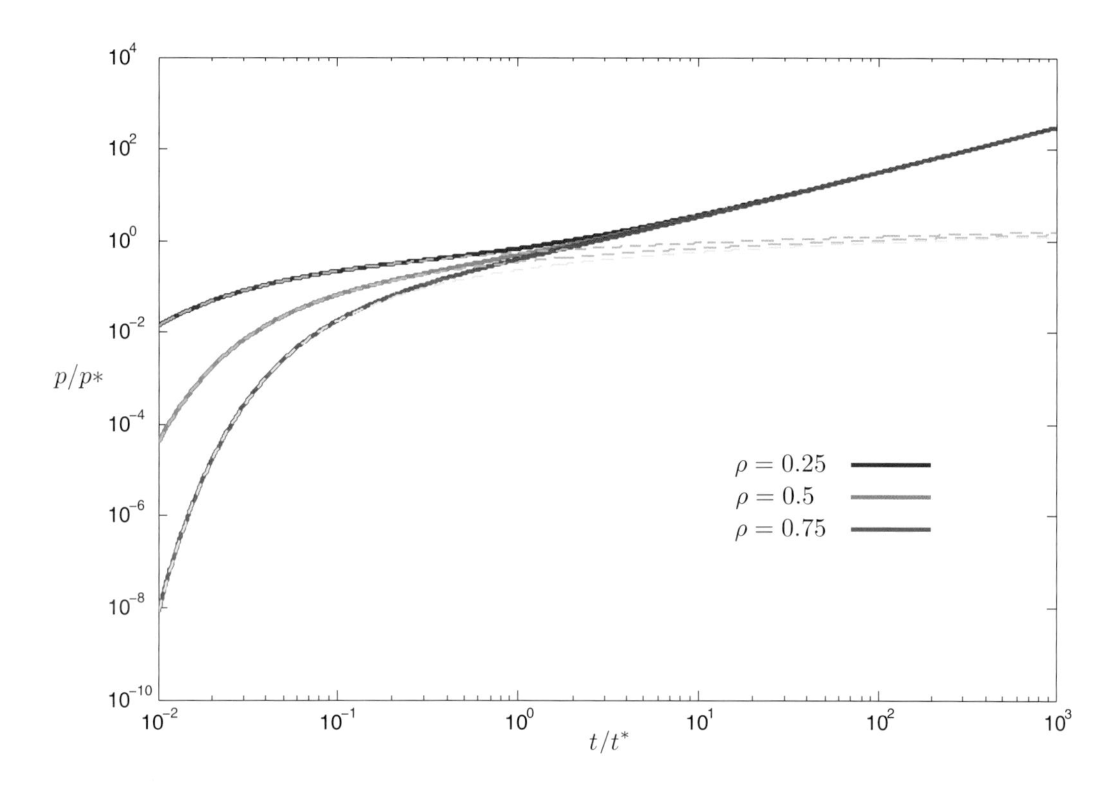

FIGURE H.1 Injection of fluid at a constant rate at the center of a disk-shaped reservoir. Plot of the dimensionless pore pressure $\Delta\rho/\rho_*$ with respect to the dimensionless time $t = t/t_*$ (equal to V/V_*) for three values of the dimensionless radius $Q = r/R$. This plot indicates that the pressure response is similar to the response of an unbounded reservoir as long as $t \leq 0.2$ and that the pressure is approximately uniform and proportional to the volume of fluid injected when $t \geq 10$. The dashed-line curves correspond to the solution **F** for an unbounded reservoir.

$$\rho_* = \mu Q_o/kL, \; t_* = L^2/c$$

with L denoting a relevant length scale of the reservoir. Also x refers to the position of the field point, and x_i to the position of the source i. At large time, the induced pore pressure is approximately given by

$$\rho \cong V/(SV_{reservoir})$$

where V is the total volume of fluid injected ($V = nQ_o t$) and $V_{reservoir}$ is the volume of the reservoir.

REFERENCES

Nicholson, C., and R.L. Wesson. 1990. Earthquake Hazard Associated with Deep Well Injection: A Report to the U.S. Environmental Protection Agency. U.S. Geological Survey Bulletin 1951, 74 pp.

Wang, H.F. 2000. *Theory of Linear Poroelasticity with Applications to Geomechanics and Hydrogeology*. Princeton, NJ: Princeton University Press.

Hydraulic Fracture Microseismic Monitoring

During a hydraulic fracture operation, very small earthquakes (**M** -4 to 0) (microseismic events) are induced from the high-pressure injection of fluids into the subsurface. These "microearthquakes" are thought to be caused by the increase in pore pressure leaking off into rock surrounding the hydraulic fracture. The increased pore pressure causes small natural fractures in the formation to slip, causing microearthquakes. These microearthquakes are thousands of times smaller than a typical earthquake that can be felt by humans. Recording and location analysis of microseismicity requires specialized seismic sensing equipment and processing algorithms. The location and size of the microseismicity are used by oil and gas operators to help determine the geometry of hydraulic fractures in the formation. Microseismic mapping is a very useful tool in planning fieldwide well development programs, such as horizontal well direction and the spacing between wells, as well as aiding the design of hydraulic fracturing procedures, such as injection rate and fluid volume. Microseismic data are acquired with either an array of seismic instruments (geophones or accelerometers) in one or multiple wellbores, or with a large number (100 to more than 1,000) of geophones near or on the surface (Figure I.1). Specialized data processing techniques are used to precisely locate the microseismic events in time and space and to compute source parameters such as seismic moment, magnitude, and moment tensors, if the data are adequate.

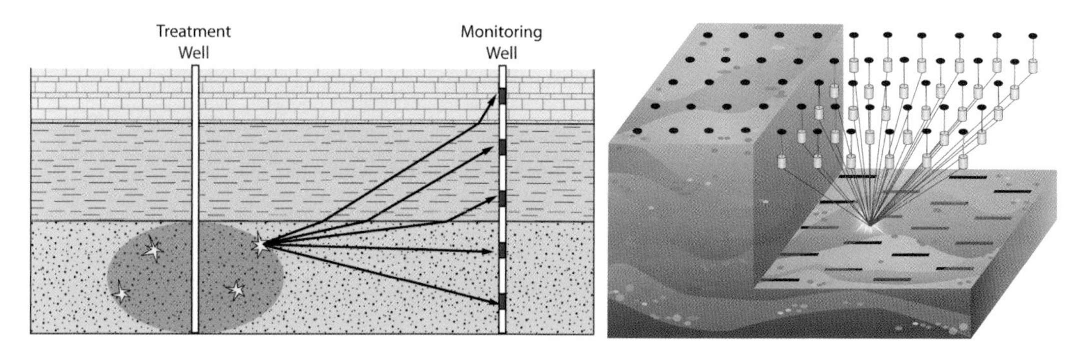

FIGURE I.1 Diagram demonstrating microseismic monitoring of a hydraulic fracture. The hydraulic fracture induces microearthquakes that are recorded with seismometers in a nearby well bore (left) or a large number of seismometer instruments placed on or near the surface (right). SOURCE: Left, courtesy MEQ Geo Inc.; right, courtesy of MicroSeismic, Inc.

The hydraulic fractures typically propagate parallel to the maximum stress direction in the reservoir. In areas of low stress differences, the hydraulic fracture pattern can be quite complex, as there is no preferential direction for the fracture to grow, in contrast with areas of high stresses, where the hydraulic fracture grows parallel to the maximum stress direction. Figure I.2 shows two examples of microseismic mapping results following hydraulic fracturing procedures in Texas: an example from the Barnett shale gas horizontal well showing a complex fracture geometry (right), and the other from tight gas sands in a vertical well in the Cotton Valley formation, which shows a simple fracture geometry (left).

Microseismic mapping with borehole or surface sensors can be used to distinguish between reactivated natural faulting and hydraulic fracture events, through **b** value analysis (see Appendix D). Hydraulic fracture wells are often drilled to avoid large natural faults distinguished from three-dimensional surface seismic images, as faults can "steal" fracturing fluid and divert fluids away from the formation targeted for hydraulic fracturing. An example of this issue was discussed by Wessels et al. (2011), where a through-going fault was reactivated during hydraulic fracturing (Figure I.3).

REFERENCES

Maxwell, S.C., J. Rutledge, R. Jones, and M. Fehler. 2010. Petroleum reservoir characterization using downhole microseismic monitoring. *Geophysics* 75(5):75A129-75A137.

Warpinski, N.R., R.C. Kramm, J.R. Heinze, and C.K. Waltman. 2005. Comparison of single- and dual-array microseismic mapping techniques in the Barnett Shale. Presented at the Society of Petroleum Engineers Annual Technical Conference and Exhibition, Dallas, TX, October 9-12.

Wessels, S.A., A. De La Pena, M. Kratz, S. Williams-Stroud, and T. Jbeili. 2011. Identifying faults and fractures in unconventional reservoirs through microseismic monitoring. *First Break* 29(7):99-104.

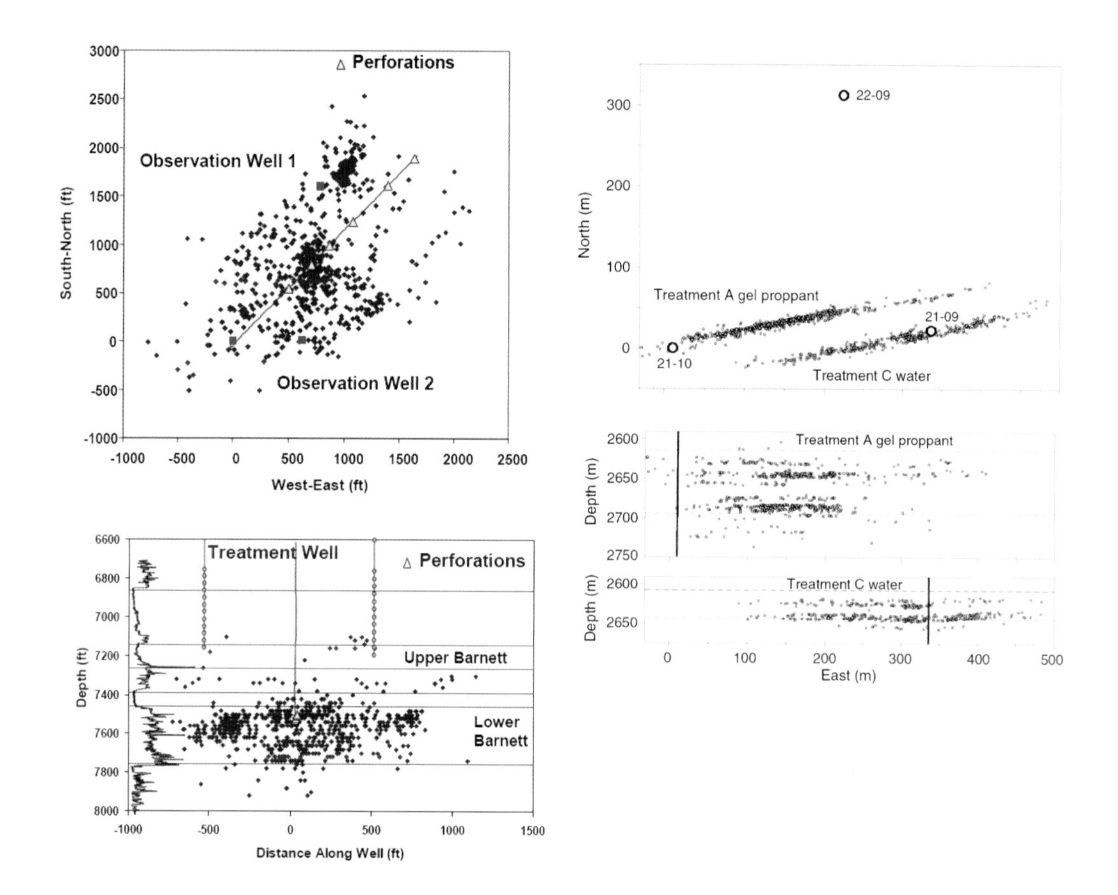

FIGURE I.2 Examples of microseismic borehole monitoring results following hydraulic fracturing procedure. (a) On the left is a map (top) and cross section (bottom) view in the Barnett Shale after a multistage hydraulic fracture treatment in a horizontal well (red line, triangles indicate perforation in wellbore where fluid is injected); the small blue dots show the location of microseismic events mapped from two borehole observation wells shown by red squares; seismic instruments are indicated by green circles. (b) On the right is a map (top) and two cross-section (bottom) views of two vertical hydraulic fractured wells (white circles) drilled in the tight gas sands of the Cotton Valley Formation. The small gray dots show microseismic locations during a gel-based and water-based hydraulic fracturing fluid injection. SOURCE: Left, Warpinski et al. (2005); right, Maxwell et al. (2010).

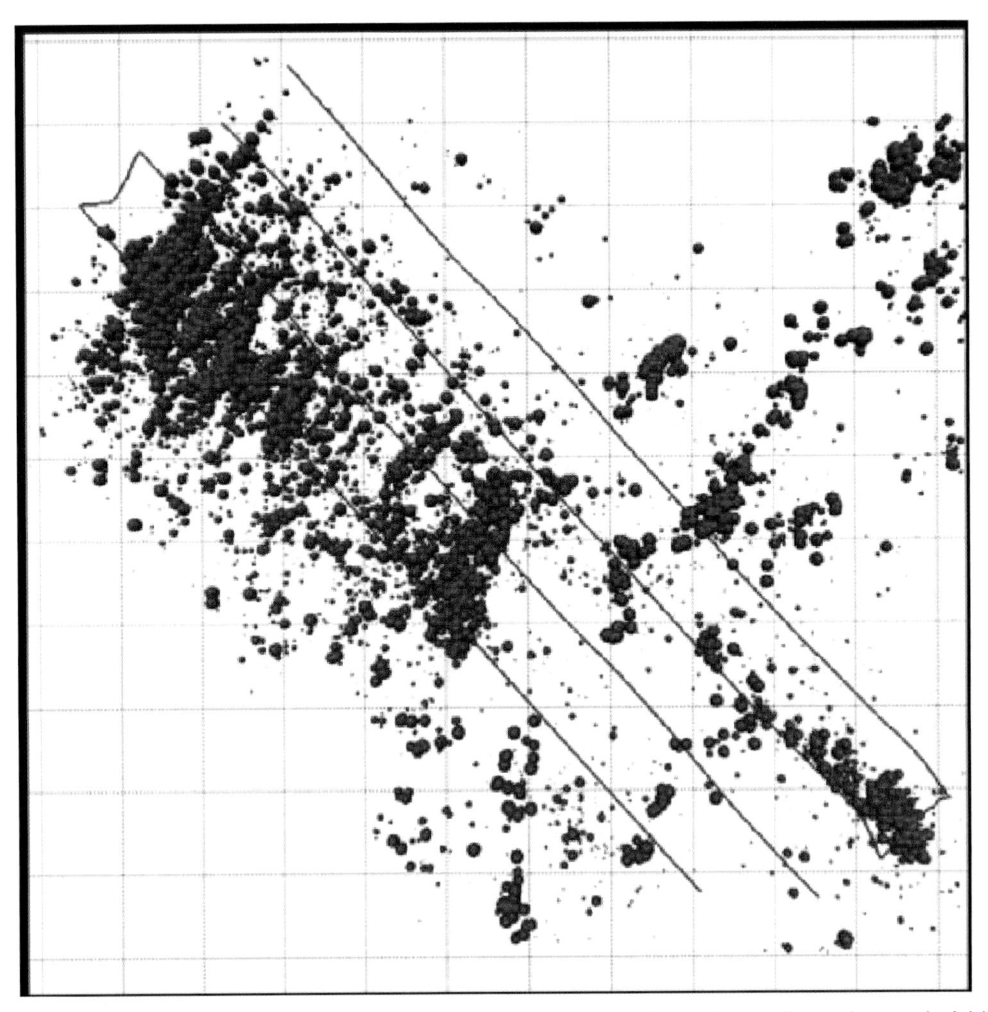

FIGURE I.3 Map view of hydraulic fracture microseismic events during a four-well stimulation (dark blue lines on the map) in the Barnett Shale. Red events are interpreted to be associated with hydraulic fracturing; blue dots indicate microseismicity associated with the reactivation of a strike-slip fault. See Wessels et al. (2011) for details. Some hydraulic fracture stages were not mapped. SOURCE: Wessels et al. (2011).

Hydraulic Fracturing in Eola Field, Garvin County, Oklahoma, and Potential Link to Induced Seismicity

A hydraulic fracture treatment in January 2011 in Eola field, Oklahoma, coincided with a series of earthquakes. Eola field is located in central Oklahoma, southwest of Oklahoma City (Figure J.1). Felt seismicity was reported on the evening of January 18 from one resident near Elmore City, Oklahoma. Further analysis showed 50 earthquakes occurred that evening, 43 of which were large enough to be located, ranging in magnitude from **M** 1.0 to **M** 2.8. The earthquakes are coincident in location and timing with a hydraulic fracture in the Eola field, Picket Unit B well 4-18. The events all occurred within 24 hours of the first activity. The deepest hydraulic fracture in the Picket Unit B well 4-18 occurred 7 hours before the first earthquake was detected. Most of the events appear to be about 3.5 km (2.2 miles) from the hydraulic fracture well (Figure J.2).

Accurate event locations were difficult to establish; the closest seismic station was 35 km (22 miles) away from the locus of the events. Errors in location are estimated to be 100-500 m (~100 to more than 500 yards) in ground distance and twice that for depth. The hypocenter depths are approximately 1 to 5 km in depth, similar to the injection depth for the 4-18 well (Figure J.3).

Other cases of suspected induced activity in Oklahoma have been reported in the past. For example, in June 1978, 70 earthquakes occurred in 6.2 hours in Garvin County after a hydraulic fracture treatment. In May 1979, a well was stimulated over a 4-day period, where three different formations were hydraulically fractured over at depths of 3.7, 3.4, and 3.0 km (2.2 to 1.8 miles). The first and deepest hydraulic fracture stage was followed by 50 earthquakes over the next 4 hours. The second stage was followed immediately by 40 earthquakes in 2 hours; no activity was associated with the third and shallowest hydraulic fracture (Nicholson and Wesson, 1990). The largest event in the sequence was **M** 1.9. Just two of the earthquakes were felt. The activity was 1 km (0.6 miles) away from the Wilson seismic station in Oklahoma.

South central Oklahoma has experienced historical seismicity (Figure J.4) and has been the most seismically active part of the state since 1977. A series of Earthscope Transportable Array stations were located near the events by coincidence; without these stations, a majority of the earthquakes could not be located.

FIGURE J.1 Google Earth image showing the state of Oklahoma and the location of the Eola oil field. SOURCE: Google Earth.

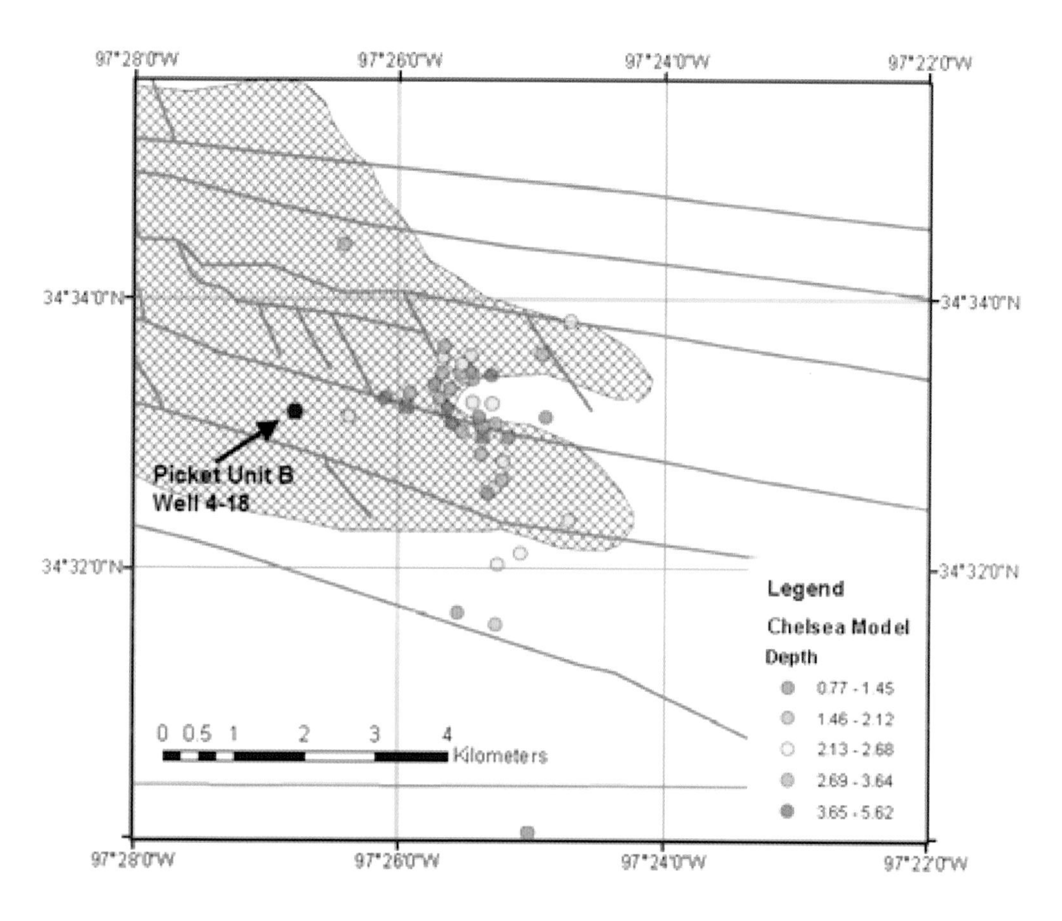

FIGURE J.2 Map of earthquake locations, the Picket Unit B Well 4-18. The Eola field is outlined by the gray hashed area. Faults mapped by Harlton (1964) are marked by green lines. SOURCE: Holland (2011).

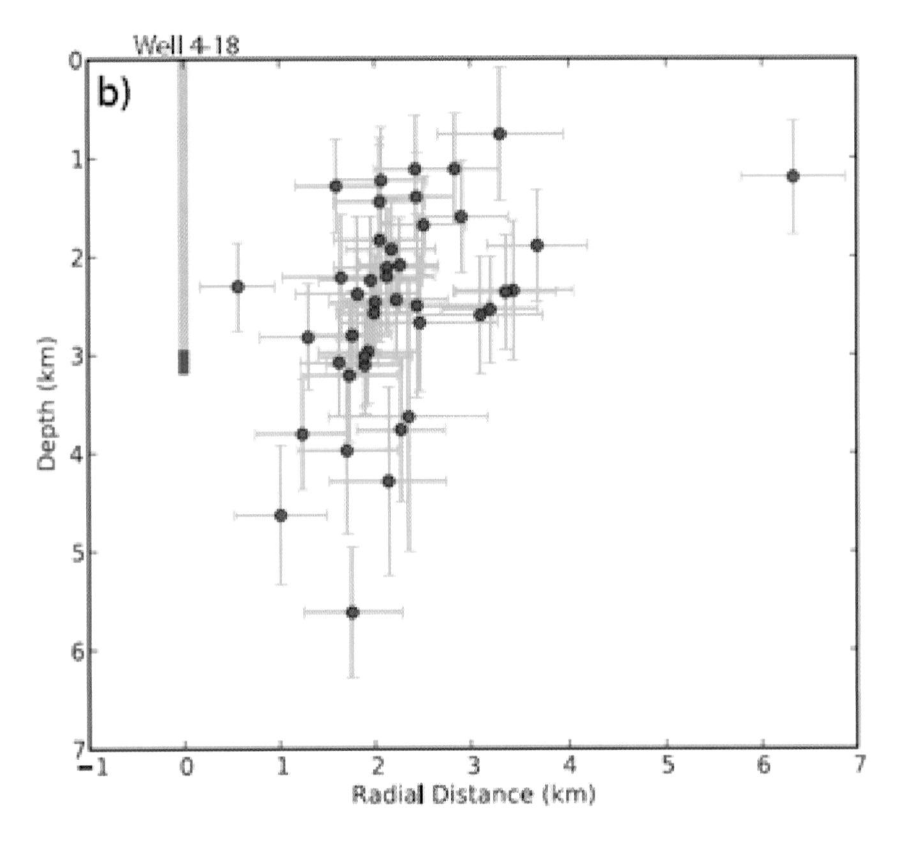

FIGURE J.3 Depth distribution of hypocenters and uncertainty estimates with respect to the fracture well 4.18. SOURCE: Holland (2011).

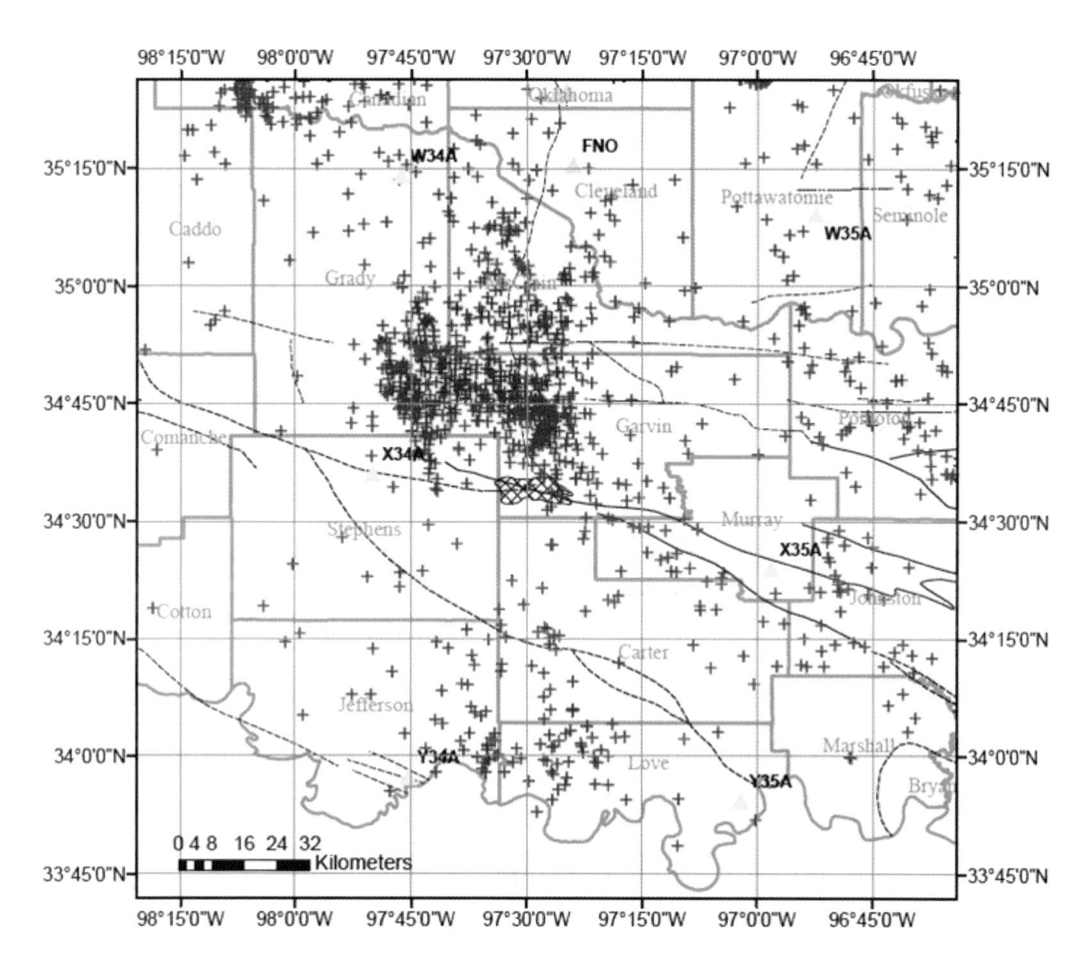

FIGURE J.4 Map of historical seismicity from the Oklahoma Geological Survey catalog. Earthquakes from 1897 to 2010 are shown by red crosses. SOURCE: Holland (2011).

REFERENCES

Harlton, B.H. 1964. Tectonic framework of Eola and Southeast Hoover oil fields and West Timbered Hills area, Garvin and Murray counties, Oklahoma. *Bulletin of the American Association of Petroleum Geologists* 48(9):1555-1567.

Holland, A. 2011. Examination of possibly induced seismicity from hydraulic fracturing in the Eola Field, Garvin County, Oklahoma. Oklahoma Geological Survey Open-File Report OF1-2011. Available at www.ogs.ou.edu/pubsscanned/openfile/OF1_2011.pdf. Accessed April 2012.

Nicholson, C., and R.L. Wesson. 1990. Earthquake Hazard Associated with Deep Well Injection—A Report to the U.S. Environmental Protection Agency. U.S. Geological Survey Bulletin 1951, 74 pp.

Paradox Valley Unit Saltwater Injection Project

The Colorado River Basin Salinity Control Project is located in Montrose County, on the western border of Colorado. The project diverts naturally occurring seepage of salt brine that would normally flow into the Delores River (and then into the Colorado River) and injects the brine underground. The project is operated by the U.S. Department of the Interior, Bureau of Reclamation. Due to concerns of induced seismicity, seismic data for this project have been continuously recorded and analyzed since the project began in 1996 in order to understand and mitigate the effects of any induced seismic events.

The Paradox Valley Unit (PVU) is a group of wells that are part of this project. The brine is produced from nine extraction wells before it can flow into the Delores River. The brine is then injected into one disposal well. The well is located near the town of Bedrock, Colorado, approximately 1 mile southwest of the extraction wells. The well injects the brine into a limestone formation at a depth of approximately 14,100 to 15,750 feet. The project began in July 1996 with an initial injection rate of 345 gallons per minute at a pressure of 4,900 psi. Current injection rates are approximately 230 gallons per minutes at a pressure of 5,300 psi.

The possibility of induced seismicity was addressed during the planning stages of the PVU injection program because the Paradox Valley Unit injection program was comparable to both the injection programs at the Rocky Mountain Arsenal northeast of Denver and the water injection program for improved oil recovery at Rangely, Colorado. Eight years before injection was begun at the PVU site, the Bureau of Reclamation commissioned a seismic monitoring network to measure the seismic activity in the Paradox Valley region. The original network consisted of 10 seismic monitoring stations. The system was upgraded to 16 stations after the injection began in 1996 and currently totals 20 stations.

Earthquakes were recorded almost immediately after the beginning of injection in July 1996 with the first seismic event measured in November 1996. Minor earthquakes continued through mid-1999, and two magnitude 3.5 events occurred in June and July 1999. In response to the higher-magnitude earthquakes, the Bureau of Reclamation initiated a program to cease injection for 20 days every 6 months. Prior to these events they had noted the rate of seismicity had decreased during the shutdowns following unscheduled maintenance. The Bureau of Reclamation hoped stopping injection twice yearly would allow time for the injection fluid to diffuse from the pressurized fractures into the rock matrix.

After a magnitude 4.3 earthquake occurred in May 2000, PVU stopped injection for 28 days to allow evaluation of the injection program and its relationship to induced seismic events. After analysis the injection rate was decreased by one-third from 345 gallons per minute to 230 gallons per minute. The program of ceasing injection for 20 days twice per year was also continued from June 2000 to January 2002 as were the lower injection rates.

In January 2002 the injection fluid was changed to 100 percent brine water from a mixture of 70 percent brine with 30 percent freshwater, which was the injection mixture from the start of the project. This heavier fluid increased the hydrostatic pressure measured at the bottom of the injection well but no difference in the rate of induced seismicity resulted from this change.

After monitoring injection into the Paradox Valley Unit injection well for almost 15 years, the Bureau of Reclamation has recorded over 4,600 induced seismic events. The largest seismic event occurred on May 27, 2000, and had a magnitude of 4.3 (see Figure K.1). After reviewing data on injection volume, injection rate, downhole pressure, and percent of days injecting, the Bureau of Reclamation noted, "Of the four injection parameters investigated, the downhole pressure exhibits the best correlation with the occurrence of near-well seismicity over time" (Bureau of Reclamation, 2009). The Bureau of Reclamation also noted the record of seismic activity appears to be divided into three distinct clusters occurring from 1997 to January 2000, 2003 to 2005, and July 2008 to the present. The Bureau of Reclamation concludes, "There appears to be a gross correlation between the three periods of increased near-well seismic activity and periods of increased time-averaged injection pressures" (Bureau of Reclamation, 2010). These conclusions reiterate the results of other investigations into the cause of induced seismicity initiated by underground injection.

The Bureau of Reclamation continues to inject saline fluids underground as part of the Colorado River Basin Salinity Control Project, and it continues to control induced seismicity by the biennial shutdown of injection activity and by limiting the volume of fluid injected. Both of these actions minimize downhole injection pressure in an effort to limit induced seismic events.

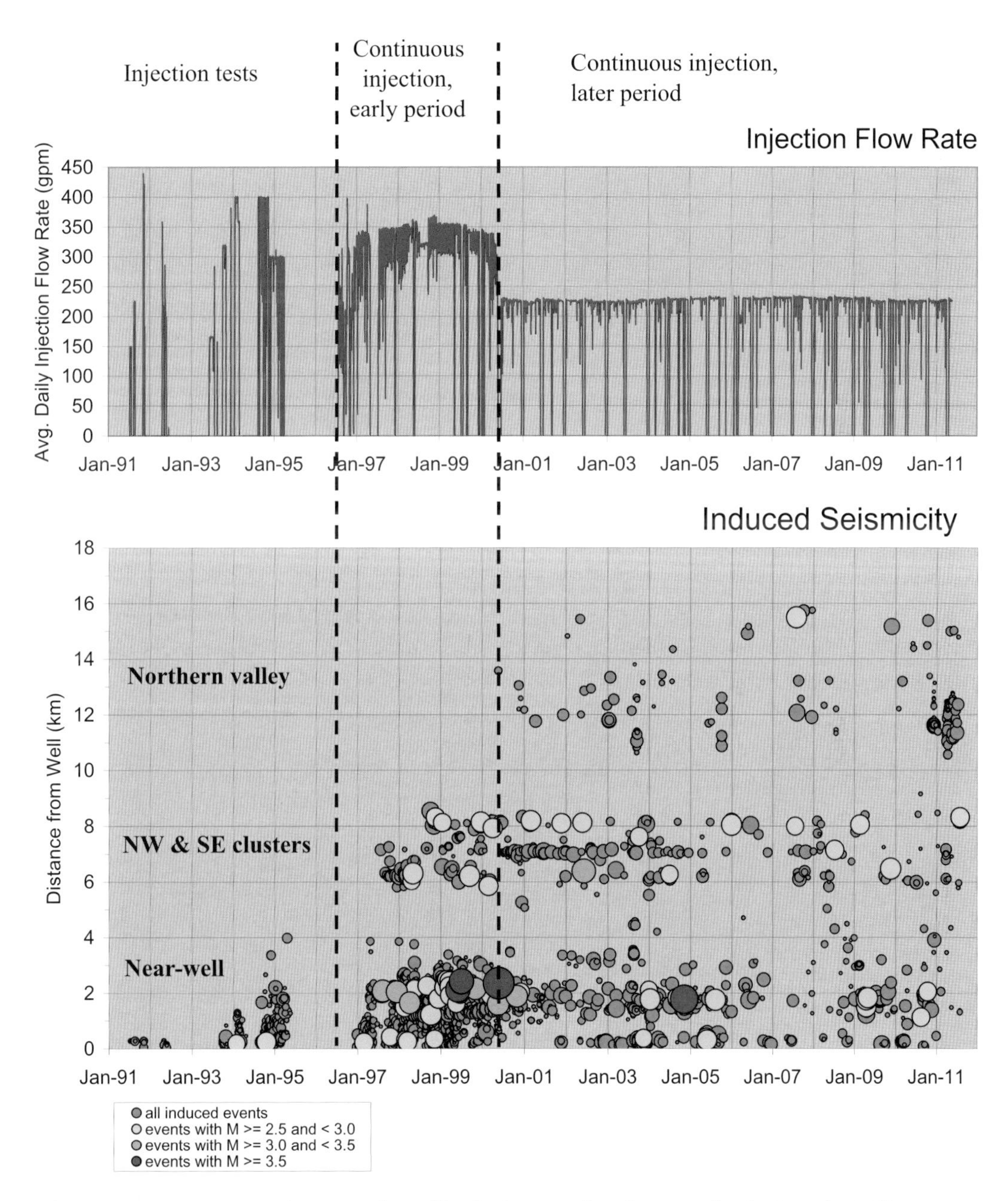

FIGURE K.1 Twenty-year data set collected by the Bureau of Reclamation for the Paradox Valley project. Upper figure shows the average daily injection flow rate in gallons per minute. Lower figure shows all induced events and their magnitudes over the same period with distance from the injection well. SOURCE: Block (2011).

REFERENCES

Block, L. 2011. Paradox Valley Deep Disposal Well and Induced Seismicity. Presentation to the National Research Council Committee on Induced Seismicity Potential in Energy Technologies, Dallas, TX, September 14.

Bureau of Reclamation. 2009. Overview of PVU-Induced Seismicity from 1996 to 2009 and implications for Future Injection Operations. Technical Memorandum No. 86-68330-2009-22.

Bureau of Reclamation. 2010. 2009 Annual Report Paradox Valley Seismic Network, Paradox Valley Project, Colorado. Technical Memorandum No. 86-68330-2010-07.

Estimated Injected Fluid Volumes

Tables L.1–L.5 contain the data used to create Figure 3.16.

TABLE L.1 Hydraulic Fracturing Volumes

Development Area	Average Volume Water (gal)	Volume Water Use Per Well (gal)	Volume Water Use Per Well (m^3)
Barnett	4,600,000	2,800,224	10,600
Eagle Ford	5,000,000	4,253,170	16,100
Haynesville	5,000,000	5,679,699	21,500
Marcellus	5,600,000	No data	No data
Niobrara	3,000,000	No data	No data
Average volume per well per day	4,640,000	—	—

NOTE: "Daily" hydraulic fracture volume plotted assumes the hydraulic fracturing procedure would take 2 days to complete; the 1-day volume plotted is half the total well volume estimated by King (2012). "Yearly" hydraulic fracture volume assumes 15 wells per year in the development area. Postfracturing flowback volume is assumed to be 20 percent of the total volume injected.
SOURCE: King (2012); Nicot and Scanlon (2012).

TABLE L.2 Carbon Capture and Sequestration Volumes

43 lb/ft^3	Density of liquid CO_2 at 80°C (AIRCO value)
2000 lb	1 ton liquid CO_2
47 ft^3	1 ton liquid CO_2 at 80°C
47,000,000 ft^3	1 million tons liquid CO_2 at 80°C per year
1,330,892 m^3	1 million tons liquid CO_2 at 80°C per year
351,355,488 gal	1 million tons liquid CO_2 at 80°C per year

Result:
1.33×10^6 m^3/year liquid CO_2 at 80°C per year
3.65×10^3 m^3/day liquid CO_2 at 80°C per year
3.51×10^8 gal/year liquid CO_2 at 80°C per year
9.63×10^5 gal/day liquid CO_2 at 80°C per year

NOTE: Table assumes 1 million tons of liquid CO_2 injection per year. The density/unit weight of liquid CO_2 varies significantly with temperature; the density of supercritical (liquid) CO_2 ranges from 0.60 to 0.75 g/cm^3 (Sminchak and Gupta, 2003). If one assumes approximately 43 lb/ft^3 (AIGA, 2009) for the unit weight of CO_2 (approximately 0.64 g/cm^3) at a subsurface temperature of 80°C (AIGA, 2009) then 1 ton of CO_2 equates to 47 ft^3, and 1 million tons/year equates to 47,000,000 ft^3/year or 1,330,892 m^3/year or 3646 m^3/day.
SOURCE: Sminchak and Gupta (2003); AIGA (2009).

TABLE L.3 Water Disposal Well Volume Calculations

9,000	bbl/day
42	gal/barrel
378,000	gal/day
137,970,000	gal/year

NOTE: Reported average saltwater disposal (SWD) injection of 8,000–11,000 bbl/day. SWD injection volumes estimated from Texas Railroad Commission for SWD wells north of DFW airport. Frohlich et al. (2010) report a survey of SWD wells in Tarrant and Johnson counties that reported rates ranging from 100,000 to 500,000 barrels per month; 9,000 bbl/day was used for graph. Nicot and Scanlon (2012) state Texas is the top shale producer in the United States.
SOURCE: Frohlich et al. (2010).

TABLE L.4 Geysers Geothermal Field Calculations

1,000,000,000	billion pounds steam/year
8	pounds steam/gallon
328,899	gal/day
120,048,019	gal/year

SOURCE: Smith et al. (2000).

TABLE L.5 Enhanced Geothermal Systems (EGS) Main Stimulation Calculations

11,500	m^3 water injected over 6 days
3,037,979	gallons water injected over 6 days
1,917	avg. m^3/day
506,330	avg. gal/day

SOURCE: Asanuma et al. (2008).

REFERENCES

AIGA (Asia Industrial Gases Association). 2009. *Carbon Dioxide*, 7th ed. Singapore: AIGA 068/10. Available at www.asiaiga.org/docs/AIGA%20068_10%20Carbon%20Dioxide_reformated%20Jan%2012.pdf (accessed May 2012).

Asanuma, H., Y. Kumano, H. Niitsuma, U. Schanz, and M. Häring. 2008. Interpretation of reservoir structure from super-resolution mapping of microseismic multiplets from stimulation at Basel, Switzerland in 2006. *GRC Transactions* 32:65-70.

Frohlich, C., C. Hayward, B. Stump, and E. Potter. 2010. The Dallas-Fort Worth earthquake sequence: October 2008-May 2009. *Bulletin of the Seismological Society of America* 101(1):327-340.

King, G.E. 2012. Hydraulic Fracturing 101: What every representative, environmentalist, regulator, reporter, investor, university researcher, neighbor, and engineer should know about estimating frac risk and improving frac performance in unconventional gas and oil wells. Paper SPE 152596 presented to the Society of Petroleum Engineers (SPE) Hydraulic Fracturing Technology Conference, The Woodlands, TX, February 6-8.

Nicot, J.-P., and B.R. Scanlon. 2012. Water use for shale-gas production in Texas, U.S. *Environmental Science and Technology* 46:3580-3586.

Sminchak, J., and N. Gupta. 2003. Aspects of induced seismic activity and deep-well sequestration of carbon dioxide. *Environmental Geosciences* 10(2):81-89.

Smith, J.L.B., J.J. Beall, and M.A. Stark. 2000. Induced seismicity in the SE Geysers Field, California, USA. Presented at the World Geothermal Congress, Kyushu-Tohoku, Japan, May 28-June 10.

Additional Acknowledgments

The committee gratefully acknowledges the support of three standing committees under the Board on Earth Sciences and Resources: the Committee on Earth Resources, the Committee on Geological and Geotechnical Engineering, and the Committee on Seismology and Geodynamics. In particular, the committee would like to thank the following people:

Committee on Earth Resources

Clayton R. Nichols, *Chair*
James A. Brierley
Elaine T. Cullen
Gonzalo Enciso
Michelle Michot Foss
Donald Juckett
Ann S. Maest
Leland L. "Roy" Mink
Mary M. Poulton
Arthur W. Ray
Norman H. Sleep
Richard J. Sweigard

Committee on Geological and Geotechnical Engineering

Edward Kavazanjian, Jr., *Chair*
John T. Christian
Patricia Culligan
Conrad W. Felice
Deborah J. Goodings
Murray W. Hitzman
James R. Rice
J. Carlos Santamarina

Committee on Seismology and Geodynamics

David T. Sandwell, *Chair*
Michael E. Wysession, *Vice-Chair*

J. Ramon Arrowsmith
Emily E. Brodsky
James L. Davis
Stuart Nishenko
Peter Olson
Nancy L. Ross
Charlotte A. Rowe
Brian W. Stump
Aaron A. Velasco